# Intensive Aquaculture

# Intensive Aquaculture

**Bruno Augusto Amato Borges**

Delve
PUBLISHING
www.delvepublishing.com

**Intensive Aquaculture**

*Bruno Augusto Amato Borges*

**Delve Publishing**
224 Shoreacres Road
Burlington, ON L7L 2H2
Canada
www.delvepublishing.com
Email: orders@arclereducation.com



ISBN: 978-1-77469-130-4 (Hardcover)

# ABOUT THE AUTHOR

**Bruno Augusto** is Aquaculture Engineer and Master on Aquaculture and Fishery Recources from Federal University of Santa Catarina (UFSC). His focus is on writing projects and development of new technologies in aquaculture. He is currently project service provider, consultant and develops research on Biofloc Technology (BFT) cultivation for fish and shrimp.

# TABLE OF CONTENTS

*List of Figures* ........ *xi*
*Abstract* ........ *xix*
*Preface* ........ *xxi*

**Chapter 1 Aquaculture Production and the Number of Cultivated Species ........ 1**
1.1 Introduction ........ 2
1.2 Fish Farming ........ 2
1.3 Crustacean Farming ........ 16
1.4 Mollusk Farming ........ 19
1.5 Alga Culture ........ 22

**Chapter 2 Fish Stocking ........ 29**
2.1 Introduction ........ 30
2.2 Scale of Protection and Assistance in Mariculture ........ 31
2.3 Fish Stocking ........ 31
2.4 Production Steps in Fish Stocking ........ 34
2.5 History ........ 39
2.6 Stocking Today ........ 41
2.7 Overfishing ........ 42
2.8 Benefits if Fish Stocking ........ 44
2.9 Sea/Ocean Ranching ........ 50
2.10 Valliculture ........ 53
2.11 Floodplain Fisheries Management ........ 56
2.12 Brush Parks or Acadjas ........ 57

**Chapter 3 Capture of Juveniles in the Wild and Growing Out in Captivity ........ 63**
3.1 Introduction ........ 64
3.2 Economic and Social Impacts of Capturing Juveniles from the Wild ... 67

3.3 Selection of Species .......... 74
3.4 Growing Out of Juveniles in Captivity .......... 78

**Chapter 4 Advantages and Weaknesses of Intensive Aquaculture Production Systems .......... 89**
4.1 Introduction .......... 90
4.2. Advantages of Intensive Aquaculture Production Systems .......... 90
4.3 Disadvantages of Intensive Aquaculture Systems .......... 104

**Chapter 5 Developments in Intensive Aquaculture .......... 113**
5.1 Introduction .......... 114
5.2 Why Intensive Aquaculture .......... 118
5.3 Difficulties of Intensive Aquaculture .......... 120
5.4 Commercial Aquaculture Production Technology .......... 122
5.5 Development of Controlled Intensive Aquaculture with Limited Water Exchange and Adjusted Carbon to Nitrogen Ratio .......... 130
5.6 Development of Sludge Remover for Intensive Aquaculture .......... 131

**Chapter 6 Feed Management in Intensive Aquaculture .......... 135**
6.1 Synoptic Review of Farm-Made and Commercial Feed Use .......... 137
6.2 Feed Production and Handling: Feed Formulation .......... 140
6.3 Manufacturing Technologies .......... 143
6.4 Feed Transport, Storage and Handling .......... 144
6.5 Top Dressing and Feed Additives .......... 145
6.6 Optimizing Feed Management Strategies .......... 146
6.7 The Role of Natural Productivity and the Implications For Feed Management .......... 150
6.8 The Use And Efficacy of Feeding Devices .......... 151
6.9 Feed Management and the Environment .......... 153
6.10 The Economics of Feed Management .......... 154
6.11 Aquafeed Policy, Regulations and Governance .......... 157
6.12 Training and Developing Information Networks to Improve Feed Management .......... 159

**Chapter 7 Role of the Government In Intensive Aquaculture .......... 161**
7.1 Introduction .......... 162
7.2 The History of Aquaculture Regulatory Framework .......... 164
7.3 Establishing an Institutional Framework .......... 167

7.4 Licensing Structures ........ 169
7.5 Access to Land ........ 170
7.6 Access to Water Resources ........ 172
7.7 Aquaculture Environmental Management ........ 172
7.8 Site Selection and Environmental Impact Assessment (EIA) ........ 174
7.9 Effluent Regulation ........ 175
7.10 Waste and Waste Management ........ 177
7.11 Coastal Aquaculture as Part of Integrated Coastal Management ........ 177
7.12 Escapement ........ 179
7.13 Regulations of Trans-Boundary Aquaculture Ecosystems ........ 180
7.14 Trade and Health Issues ........ 180
7.15 Introduction and Movement of Species ........ 181
7.16 Food Safety and Health ........ 181
7.17 Mitigating Investment Risks ........ 182
7.18 Aquaculture and Zoning ........ 183
7.19 Simplification of Administrative Processes: Single Window Approach ........ 185
7.20 Financial Incentives ........ 186
7.21 Social Impacts ........ 187
7.22 Conclusion ........ 188

**Chapter 8 Global Trends in Intensive Aquaculture ........ 191**
8.1 Introduction ........ 192
8.2 Factors for Aquaculture Development ........ 193
8.3 Production Relating to Inputted Energy ........ 196
8.4 Tackling the Obstacles to the Development of Aquaculture ........ 198

**Chapter 9 Factors Affecting Intensive Aquaculture Production ........ 207**
9.1 Introduction ........ 208
9.2 Physical Factors Influencing Intensive Aquaculture ........ 208
9.3 Markets ........ 216
9.4 Management ........ 218
9.5 Political Goodwill ........ 219
9.6 Society ........ 220

**Chapter 10 Aquaculture Products and Marketing** .... **221**
10.1 Introduction .... 222
10.2 Aquaculture Products .... 225
10.3 Crustaceans .... 225
10.4 Finfish .... 230
10.5 Aquaculture Marketing .... 236
10.6 Benefits of Aquaculture Marketing Products .... 238
10.7 Features of Marketing of Aquaculture Products .... 238
10.8 Marketing Channels .... 239

**References** .... **241**

**Index** .... **247**

# LIST OF FIGURES

**Figure 1.1**: Tilapia farming is quite easy. These fish are quite hardy and are very easy to grow. They are also in high demand not only at homes but also in restaurants and seafood outlets.

**Figure 1.2**: In order to have a successful Tilapia business, one needs to follow all the guidelines. Having self-sustaining Tilapia system can be very profitable given the fact that the farmer needs not to outsource the fingerlings and the fingerlings will be hatched in the temperatures of the water hence no need to adjust.

**Figure 1.3**: The catfish is considered as a great alternative for the Tilapia since they multiply quite quickly and are easier to rear. They are also in high demand and are very nutritious.

**Figure 1.4**: Preparation of a catfish pond. One should consider that the ground is the farm and that they have killed the native fish that was present in the pond.

**Figure 1.5**: The salmon is highly demanded in most homes and restaurants. It can be bought as fillets or one can decide to buy the whole fish.

**Figure 1.6**: Salmon is very popular especially the Atlantic one. The farms are in crisis given the reports that they are costing the economies, the societies and the environment. The salmon can be very damaging to the environment.

**Figure 1.7**: The blue fin tuna. A common type of tuna is commonly raised and is very expensive.

**Figure 1.8**: Fishponds. These are ponds used for rearing fish.

**Figure 1.9**: Indoor fish rearing where the temperature is regulated and it is covered on all sides.

**Figure 1.10**: A prawn. A type of crustacean reared for consumption.

**Figure 1.11**: A lobster. A common delicacy in Chinese restaurants. It is raised on farms.

**Figure 1.12**: Cages that are used for raising crabs. They are floating on seawater and arranged in water for easy feeding.

**Figure 1.13**: Oysters are becoming very nutritious to the world population. They are easy to grow and mature very fast. Farmed oysters are preferred given the fact that they are cleaned and are free from contamination.

**Figure 1.14**: Farming of algae by use of agricultural technology.

**Figure 1.15**: Algae farming by use of ponds. They are quite cheaper to construct and generates the most output if the ponds are big.

**Figure 2.1**: Stocking fish in a river in California.

**Figure 2.2**: Tanks in a shrimp hatchery.

**Figure 2.3**: Juvenile salmon towards the end of their stay in a hatchery.

**Figure 2.4**: 400 tons of jack mackerel caught by a Chilean purse seiner. Overfishing catches too many fish simultaneously so that the breeding population is too weary to recover. Overfishing is typically combined with inefficient types of commercial fishing, which are subsequently thrown away into enormous quantities of undesirable fish or other animals.

**Figure 2.5**: Fishing for stocked trout in Wyoming.

**Figure 2.6**: Fish are stocked all over the world. This river in Mount Aspiring National Park, New Zealand, is stocked with trout.

**Figure 2.7**: Costs and benefits of fish stocking.

**Figure 2.8**: Fish farming is a form of aquaculture in which fish are raised in enclosures to be sold as food. It is the fastest-growing area of animal food production. Today, about half the fish consumed globally are raised in these artificial environments. Commonly farmed species include salmon, tuna, cod, trout and halibut. These "aquafarms" can take the form of mesh cages submerged in natural bodies of water or concrete enclosures on land.

**Figure 2.9**: In view of wild-caught seafood overfishing globally, ocean fishing has increased fresh fish to western markets and become one of the world's fastest-growing sectors. Precise reporting of wild caught fish quantity was hard and we asked whether in the statistics for farmed fish production comparable errors exist in data. There is widespread marine fish farming in the Mediterranean and stationary cages from the coast of 16 nations may be observed using Google Earth satellite images. Example image from Google Earth showing a fish farm off the coast of Greece.

**Figure 2.10**: In the lagoon, the engineering work also serves a range of goals such as providing greater fishing surfaces and the passage by the neighboring lagoons of fisheries and shellfish. Interconnecting waterways also provide better access to nursery regions for larval fish and shrimps. Sedimentation reduction and circulation enhancement are other aims of internal canals. Hydraulic engineering within lagoon systems is exemplified by the Huizache-Caimanero lagoon complex in Mexico.

**Figure 2.11**: In the lagoon, the engineering work also serves a range of goals such as providing greater fishing surfaces and the passage by the neighboring lagoons of fisheries and shellfish. Interconnecting waterways also provide better access to nursery regions for larval fish and shrimps. Sedimentation reduction and circulation enhancement are other aims of internal canals Alternative migration patterns of riverine 'whitefish' and 'blackfish'.

**Figure 2.12**: The country can be divided hydrologically into five principal drainage areas.

**Figure 2.13**: Aquaculture fishing.

**Figure 3.1**: Aquaculture is a multi-billion-dollar enterprise that employs a variety of fish production methods. One of these strategies is to harvest the wild people in confined and controlled conditions, from extremely early stages in their life cycle to large, mature adults.

**Figure 3.2**: Special demands on animal welfare are made by capture-based aquaculture. Therefore strategies and procedures created in regard to their gentleness ought to be documented.

**Figure 3.3**: For intervals of several hours to several months, the fish are preserved alive. Our investigations reveal that the quality is substantially greater by using short live storage on board trawlers than by slaughtering fish straight after being pulled on board.

**Figure 3.4**: The aquaculture production of cobia in pools may lead to water quality problems and excess nutrient loading in the effluent, depending on the pond, crop density, feed rates, and water exchange rate. Located on the shore, large-scale cobia production at different phases of life could affect any coastal environment and necessitate monitoring of the production outflow.

**Figure 3.5**: While it is unlikely that post-larvae capture and cultivation will meet the demand of the ornamental trade for all tropical fish, the capture and cultivation of post-larvae can create large niche markets for eco labeled specimens and provide coastal villagers sustainable economic advantages from coral reef resources.

**Figure 3.6**: Early adoption and extensive adoption of well-managed selective breeding operations, which compounded increases in growth performance and qualities such as disease resistance, were vital to this achievement for Atlantic salmon. However, genetic technology differs enormously between species and these technologies progress quickly and become more accessible and affordable.

**Figure 3.7**: CBA also affects the stressors introduced to fish during capture and the duration of these effects increases considerably from hours and minutes for traditional fishing to days and months for CBA and also in terms of adaptability to new surroundings.

**Figure 3.8**: In all stages of the Mahi aquaculture, researchers have made steady progress and agree that difficulties can be addressed if adequate resources are used. The technology is in place, but the successful renewal of efforts to build commercial aquaculture for the species is eventually determined by market forces. If there is good demand for the market and favorable prices, mahi may be increased.

**Figure 3.9**: Apart from selling totoaba in restaurants and retail outlets throughout Mexico, they have supported the restoration of wilderness from the beginning, releasing to their natural environment thousands of Totoaba juveniles and working with local schools to educate the coming generations on the significance of the conservation of wildlife. So far over 100,000 young people have been released into the wild by Earth Ocean Farms to safeguard this unique Mexican species.

**Figure 3.10**: Lower tropical farming tackles some of aquaculture's environmental issues. Some of the hurdles for the expansion of rabbitfish aquaculture involve the provision of hatchery fingerlings to replace wild fish, the creation of algal-specific diets and an improved understanding of and mitigation of illnesses.

**Figure 3.11**: The use of hatchery releases with various management targets is referred to as restoration ("to restoring depleted spawning biomass to the level where it again offers a regular substantial yield"), stock improvement ("to increase a natural supply of juveniles and optimize their harvest by eliminating recruitment restrictions").

**Figure 3.12**: Aquaculture based on capturing entails catching and then keeping wild animals alive or feeding them to harvest. The purpose is to boost biomass, enhance the quality of raw materials and the chance to act strategically with regard to the market by determining the harvest and sale of the fish. For example, a short and frantic season characterizes fishing for cod. If wild fish are harvested, and kept alive in the cage, fresh cod may not be provided in the short fishing season but throughout the year.

**Figure 4.1**: Fish production has increased rampantly due to intensive fish farming.

**Figure 4.2**: Many employment avenues have been created through intensive fish production systems.

**Figure 4.3**: Fish markets have grown due to the production of more fish.

**Figure 4.4**: Wasted lands are made useful through the construction of ponds.

**Figure 4.5**: Fish is considered a healthy source of food.

**Figure 4.6**: Fish are a good source of omega and very nutritious.

**Figure 4.7**: Water recycling systems are used in efficient water management.

**Figure 4.8**: Government earns revenue through the exportation of fish.

**Figure 4.9**: RAS is among the technologies used in intensive fish production systems.

**Figure 4.10**: There are a variety of diseases that could affect fish in intensive fish farms.

**Figure 4.11**: Some of the activities done in intensive fish farms have negative effects on the environment.

**Figure 4.12**: Variety of feeds can be used in rearing fish.

**Figure 5.1**: An intensive aquaculture system for raising tilapia.

**Figure 5.2**: An oxygen diffuser used in the addition of oxygen to water.

**Figure 5.3**: *Clarias* spp is one fish that is known to be used for intensive fish farming.

**Figure 5.4**: Organic fertilizers are normally used to help in the growth of organic foods in the fishponds.

**Figure 5.5**: Fish tank used to raise fish. The feed is given to the fish and water used is recycled after being used.

**Figure 5.6**: Tanks used, as fishponds can be disadvantageous especially if the fish are big.

**Figure 5.7**: An oxygen generator. The machine is complicated and requires a skilled technician and constant monitoring.

**Figure 5.8**: Murray cod industry development plan aquaculture water resources.

**Figure 5.9**: Hand stripping of Murray cod eggs.

**Figure 5.10**: Murray cod can be very large and is a very expensive fresh water fish.

**Figure 5.11**: Hybrid Murray cod is a technological development in the aquaculture industry.

**Figure 5.12**: Aeration of fish ponds is important so as to ensure better growth of the fish in the ponds.

**Figure 5.13**: Sludge needs to be removed from the fishponds to ensure proper development of the fish in the ponds.

**Figure 6.1**: Feed distribution using an automatic feeder for cage culture of Atlantic salmon near Bodo, Norway.

**Figure 6.2**: Feeding striped catfish (*Pangasianodon hypophthalmus*) with farm-made aquafeed, Mekong Delta, Vietnam.

**Figure 6.3**: Preparation of farm-made supplemental feed using locally available feed ingredients, Mymensingh, Bangladesh.

**Figure 6.4**: Fish nutrition, feeds, and feeding management play important roles in increasing the productivity of aquaculture farms.

**Figure 6.5**: Production of high-quality feeds starts with good quality ingredients and proper quality control throughout processing until use in fish farms.

**Figure 6.6**: A nutritionally-balanced feed and adequate feeding are important factors that maximize fish production and profitability.

**Figure 6.7**: A good quality feed and appropriate feeding management maximize feed efficiency, lower cost of production, and reduce environmental degradation.

**Figure 6.8**: Monitoring feed consumption via feed trays can indicate adjustments in feed applications that maintain water quality and enhance shrimp health.

**Figure 6.9**: Auto-feeders intermittently distribute small quantities of feed that are caught by the shrimp before sinking to the pond bottom and losing nutrients.

**Figure 6.10**: In biofloc systems, strong aeration keeps organic matter in perpetual suspension. Significant amounts of microbiota "piggyback" on the organic matter, recycling the rich nutrients available in the water.

**Figure 6.11**: Economic planning is crucial.

**Figure 6.12**: The correlation between feeding intensity, growth and FCR.

**Figure 6.13**: Fish food pellets.

**Figure 7.1**: Aquaculture farm.

**Figure 7.2**: Will the U.S. development of aquaculture technology for emerging species like pompano lead to commercial success outside the United States?

**Figure 7.3**: Aquatic food security.

**Figure 7.4**: Proportion of global aquaculture production by volume by country. Data from FAO (2015).

**Figure 7.5**: Simplified aquaculture production functions map.

**Figure 7.6**: Trends in global production from marine (red)- and freshwater (yellow)-capture fisheries and aquaculture (blue). Data from FAO FishStatJ (FAO 2015).

**Figure 7.7**: Location of aquaculture sites in the UK. Location and farm-type data from information compiled in the UK Multiannual plan for the development of sustainable aquaculture.

**Figure 7.8**: Promote sustainable commercial aquaculture.

**Figure 7.9**: Schematic model of national Nr cycling in China's aquaculture ecosystem.

**Figure 7.10**: Reducing the risk of aquatic animal disease outbreaks and improving the environmental management of coastal aquaculture in Vietnam.

**Figure 7.11**: Integrated Coastal Zone Management.

**Figure 7.12**: Alaska exceeds Canadian chinook escapement goal, decline remains a mystery.

**Figure 7.13**: Schematic figure of a designated aquaculture zone (hatched area in blue color) representing an estuary and the adjacent coastal marine area. Individual farms/ sites (F) owned by different farmers, are presented in different colors. Four clusters of farms illustrate examples of aquaculture management areas (AMAs), grouped according to a set of criteria that include risks and opportunities.

**Figure 8.1**: In 2007, a limited number of substantial countries, like Thailand, Spain, and Canada, underwent contraction of production. This was mostly due to market and competitiveness; however, sickness and unique natural events could also take part in one-year results. Overall, these losses represented 1.6% of global supplies (i.e., more than compensated by growth elsewhere).

**Figure 8.2**: The structures of the market chain are critically linked to the viable size of the company. Europe's smaller businesses tend directly to market consumers and local restaurants and hotels, earning a considerable premium at normal wholesale market rates. It is not a solution that will saturate local markets for somewhat larger companies.

**Figure 8.3**: With demand on freshwater supplies currently increasing, a further expansion in aquaculture can be projected to use more plentiful brackish and marine waters. Environmental challenges are no less difficult, however.

**Figure 8.4**: Aquaculture will proceed to utilize most of the fish meal and oil produced unless alternatives for better values continue to evolve, but it will not be enough to fulfill increasing requirements for aqua food ingredients.

**Figure 8.5**: In the same bodies of water, or by some other water-based connection, integrated multi-trophic aquaculture systems may be described as cultivation systems using species from various trophic levels (for land-based systems). The scale must not be large if the species are cultivated and the quantities produced are compatible.

**Figure 8.6**: In poor countries, integrated aquaculture/agriculture provides a mechanism to broaden and increase productivity for rural systems. The method of culture differs from the single-culture approach, which typically poses too high a risk for poor farmers. Integrated systems profit from synergy between the different components and have production variety which gives rise to relative environmental health.

**Figure 8.7**: Selective reproduction can enhance the annual performance of farmed fish stocks for a variety of features, but it is frequently still essential for these fish to use several other strategies to realize their maximum potential. The end yield can be greatly decreased when sexual maturity in production fish can become more aggressive, stop growing, lose condition and become more disease sensitive. One or another sex is preferred in many animals, e.g., because in crop sized, it grows quicker or remains immature. Females generally develop later than males in salmonids.

**Figure 8.8**: Offshore cage aquaculture has no universally accepted definition. Sites are characterized in Norway by wave height, while offshore aquaculture activities are described in the United States as activities in the unique economic zone ranging from a three-mile coastal territory boundary to 200 miles offshore.

**Figure 9.1**: The increased demand for fish all over the world needs that individuals start fish farming.Source: FOA of the UN.

**Figure 9.2**: Bacteria are spheroid or rod-shaped; they are microscopic and cannot be seen by naked eyes.Source; POND Trade Magazine.

**Figure 9.3**: Sludge formed at the bottom of a tank, discolor water when disturbed and reduce water volume in a pond.

**Figure 9.4**: Discoloration of water due to algae.

**Figure 9.5**: A lot of capital is used to set up an aquaculture farm.

**Figure 9.6**: Fish farming technologies in intensive aquaculture are known to be the future of the fish agricultural systems to improve the quantity and quality of fish.

**Figure 9.7**: Intensive fish farms need to be established in areas close to the markets because of high cases of perishability and the need for freshness.

**Figure 9.8**: An intensive fish farmer needs to be hardworking and committed to the business in order to ensure the success of the business.

**Figure 10.1**: There are a variety of aquaculture products sold in both local and international markets.

**Figure 10.2**: Developments made to the aquaculture sector have resulted in the increase in the amount of aquaculture products produced.

**Figure 10.3**: Products such as shrimps are in high demand in the aquaculture market.

**Figure 10.4**: Shrimp is used in various restaurants in preparing certain dishes.

**Figure 10.5**: Different species of crayfish are sold in international markets.

**Figure 10.6**: Different methods are used in the marketing of crabs.

**Figure 10.7**: Most aquaculture farmers opt to farm salmon as it is easier.

**Figure 10.8**: Trout production and marketing is second to that of Salmon.

**Figure 10.9**: Tilapia is the most common species of finfish aquatic animals.

# ABSTRACT

Fish farming is a type of aquaculture wherein fish are nurtured in enclosures for food. It's far the quickest developing field of animal food generation. Around half of the fish taken globally are nurtured by farming. Usually farmed species consist of tuna, salmon, cod, halibut and trout. Aquafarms often take the form of web cages immersed in water bodies such as lakes, or concrete land enclosures. The United Nations estimates that roughly 32% of worldwide fish numbers are overexploited, depleted or recuperating and must be urgently restored. Fish farming is welcomed as a strategy to prevent overfishing. However, the book mentions that these farms have a long way to go, and might severely damage ecosystems through introducing sicknesses, pollution and invasive species. The damage coming from fish farms varies, based on the kind of fish, how it's nurtured and fed, the population size, and so on. Overfishing of the smaller "forage" fish also has effects on the oceanic ecosystem.

# PREFACE

Aquaculture's proportion of the forage-fish proportion has almost doubled from 2000. It now covers almost 70 percentage of the worldwide fish meal supply, and nearly 90 percent of the sector's fish oil market. This book mentions that so popular is the market that many countries are nowadays sending ships to Antarctica, in a bid to reap thousands of heaps of krill per year. Although lots of the krill wind up in prescribed drugs and various merchandise, to critics of aquaculture the concept of depleting the lowest food chain with a view of getting highly-priced protein feels like ecological madness.

Fish farmers are getting greener, farming omnivorous fish such as tilapia and the use of feeds containing soybeans and different other grains; besides salmon feed in recent times is normally no less than 10 percent of fish meal. The amount of forage fish used in keeping with fish output has fallen by using roughly 80 percent from what it was 15 years ago, and the book mentions it could fall much further.

Whether you're nurturing fish in an offshore enclosure, or in a filtered land tank, you ought to feed them. But they have one massive advantage over land animals: you need to feed them a lot less. Fish require fewer calories, because they're cold-blooded and because, residing in a buoyant environment, they don't contend with gravity as much. It takes roughly a pound of feed to produce a pound of farmed fish; in comparison, two pounds of feed is needed to nurture a fowl, roughly three for a pound of beef, and seven for a pound of beef. Fish farming provides a constant supply of animal protein which can fulfill the needs of billions on earth—especially omnivorous fish like carp, catfish and tilapia.

The book further mentions that escaped fish poses a great risk to the environment. Annually, thousands of fish escape from farms and enter the ecosystem, where they compete with other aquatic creatures for feed. High stock numbers often cause a high level of pollution from fish waste and unconsumed food; this consequently leads to reduced stock numbers.

Outdoor fish farms also can draw predatory marine animals, consisting of sea birds and sea lions, which are every so often poisoned or hunted down by aqua farmers for consuming the fish. Despite proof to the opposite, it is still a commonplace false impression that fish can't sense pain. Slaughter techniques in the aquaculture sector are inhumane. Minimal attention is provided to the struggling animals, and most remain conscious during the duration of slaughter. Furthermore, some species, such as salmon are starved for several days to drain out the gut before slaughter. Some fish are also shocked or killed by bleeding them out, being struck on the head repeatedly, suffocating or freezing. In many countries, there are no policies to protect the gentle treatment of fish. Years of unfettered and underreported fish catches are affecting the lifestyles of rigorously overfished species, inclusive of Bluefin tuna within the Mediterranean and Atlantic. To fulfill the excessive and growing need for sushi in Asia, aquaculture is becoming a popular practice.

Chapter 1

# AQUACULTURE PRODUCTION AND THE NUMBER OF CULTIVATED SPECIES

## CONTENTS


1.1 Introduction.....2

1.2 Fish Farming.....2

1.3 Crustacean Farming.....16

1.4 Mollusk Farming.....19

1.5 Alga Culture.....22

## 1.1 INTRODUCTION

Aquaculture is also referred to aquafarming. It involves the farming of fish, crustaceans, mollusks, aquatic plants, algae, and other organisms. It involves the cultivation of fish from freshwater and saltwater sources. The populations are normally done in controlled conditions and in most cases can be contrasted with commercial fishing. Commercial fishing normally involves the fishing of the wild fish. Mariculture is the name given to the cultivation of fish in marine waters, which are normally opposed to freshwater breeding (Waagbo & Remo, 2020). According to the Food and Agriculture Organization (FAO), aquaculture involves the farming of aquatic organisms, which include fish, mollusks, crustaceans and aquatic plants. There needs to be some form of intervention and rearing process that help in enhancing the production of the same, feeding, protection from predators, regular stocking among other aspects. Some forms of aquaculture are known to be very harmful to the environment through nutrient pollution and others are known to transfer diseases to the wild populations. As such, some countries have been able to ban the production of some kinds of aquatic life. Aquacultures of different kinds are further discussed below.

## 1.2 FISH FARMING

It is usually referred to as aquaculture or pisciculture, and it involves the rearing of fish most especially for food in fish tanks or some use artificial enclosures such as fishponds. Some people are rearing fish for aesthetic value and it is now a lucrative business and some companies have even patented some of the fish. It is commonly a sub-type of aquaculture as it involves controlled cultivation and harvesting of animals that are found in the aquatic environment. The most common types of fish that are commonly reared in fish farms include; carp, catfish, tilapia and salmon (Tucker & Hargreaves, 2009).

Tilapia is a very common type of fish and is mostly used for fish farming in Africa. They have fast growth, are large and their meat has high protein content. It thrives well in temperatures ranging between $28^0$C and $30^0$C. In tilapia culture, the main challenge is controlling breeding as they breed very rapidly. If not well maintained they increase in number and this leads to stunted growth. They prefer cereals instead of feeding on other fish. These fish are referred to as warm water fish because of the temperatures they can survive in. they are mostly found in lakes and less in the oceans and seas. These fish are normally picked by the farmers of the same since they are quite easy to grow and are quite hardy. To make tilapia farming easier one needs to set up a growing system that is quite easier for them, especially one that will fit the lifestyle and one that is easy to maintain.

Tilapia are good fish to eat and it is quite easy to create new recipes within the household that are healthy and safe. The demand for fresh tilapia is always on the rise not only at homes but also in restaurants and seafood outlets. In an aquaponics system, Tilapia are normally reared alongside vegetables. In this system, there is a kind of beneficial relationship between the two. The tilapia waste normally contains nutrients that are used in vegetable farming such as the growth of lettuce, cucumbers, tomatoes and other plants. On the other hand, these vegetables help in the purification of water that is used by the fish. In order to participate in fish farming, one needs to assess their readiness and personal motives. This means that they need to have a plan of how to capitalize the business, how to feed the fish, and whether he/she needs to find a market for the fish if the motive is to sell the fish. One should also see whether they have a conducive place to set up the establishment.

**Figure 1.1**: Tilapia farming is quite easy. These fish are quite hardy and are very easy to grow. They are also in high demand not only at homes but also in restaurants and seafood outlets.

*Source: Aquatic Network*

The other aspect is that, before one starts tilapia farming, they need to consider the state regulations, which means that they should determine whether there are any regulations for obtaining and farming tilapia. One thing that should be considered by tilapia farmers is that they are invasive kinds of fish. This means that they can be able to displace the native fish populations. As such, before introducing them to the natural water bodies, one needs to be very careful of what they are doing. As such, they need to be properly displaced and one needs to be very careful when disposing of any live fish or waste that contains eggs in the water bodies (Tacon & Halwart, 2007). Another instance is that if it is in your home garden, if one disposes of fish anyhow, one may end up with compost from the fish that you do not consume within the household. Another aspect is the development of a budget and a plan of action. In any business or venture, one needs to have a plan and a budget s they are the most crucial things in running the business. If it is not too hard, then one can have a team of individuals that can be able to help them in maintaining the venture. Setting up a tilapia system is also an important aspect given the fact that the business is about Tilapia farming. There are a number of environments that the Tilapia can be grown successfully. They include ponds, cages, raceways, and tanks. Reports show that, in the urban centers, there are those that grow these kinds of fish in trashcans. Growing of the fish in the ponds is so far the common type is perhaps considered the simplest method. For the case of using a tank or a cage, the farmer needs to be able to purchase the right materials that can be used to build the cages and tanks. In the case where the water is not being circulated, the farmer needs to find equipment for reconditioning the water before the introduction of fish. After being able to figure out all the mentioned aspects, it is ideal to acquire the fish to use in the ponds. This should be done carefully and as such, one needs to find a reputable dealer to sell the tilapia to them because of the fact that one would need to avoid having eggs from other kinds of fish mixed up with the tilapia. To make the process successful the farmer needs to acclimate the fingerlings immediately after receive so that they can be able to adjust to the current temperatures of the water.

**Figure 1.2**: In order to have a successful Tilapia business, one needs to follow all the guidelines. Having a self-sustaining tilapia system can be very profitable given the fact that the farmer needs not to outsource the fingerlings and the fingerlings will be hatched in the temperatures of the water hence no need to adjust.

*Source: Self-sustaining ecosystem*

After all this is done, the farmer needs to be patient with the fish and be very discipline. This means that he/she needs not to harvest at any moment they want to but needs to grow them until they reach harvestable sizes. It can be quite tempting especially once one sees that there is one big healthy tilapia. As such, one needs to be very careful. One should also know to always maintain the temperature of the water, and also free the ammonia brought about by the waste of the tilapia. Just like any agricultural practice, especially that of vegetables, the farmer needs to be very careful and provide proper care and maintenance to the fish (Sivasankar et al., 2017). To promote optimal growth of the fish, the farmer needs to also look at the aquacrop. If they are maintained in the correct manner, a farmer can be able to harvest the tilapia after 8 months. During harvest, one should be smart not to harvest all the tilapia. One needs to leave some of the adults to be able to breed and produce fry and fingerlings that will be able to maintain the fish crop for the next harvest. This will make the farmer have a self-sustaining business instead of depending on other dealers for tilapia fingerlings.

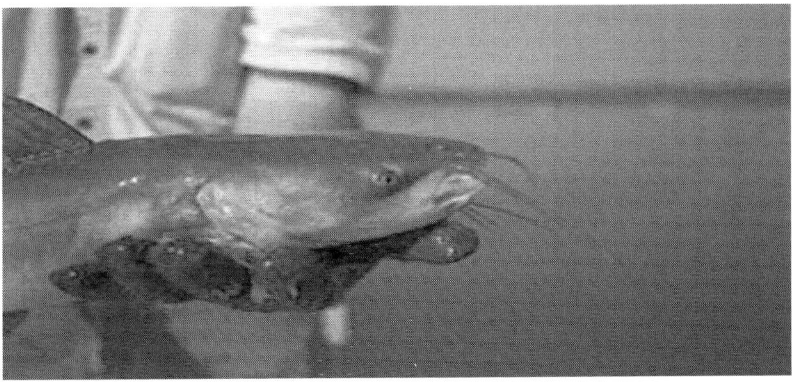

**Figure 1.3**: The catfish is considered as a great alternative for the Tilapia since they multiply quite quickly and are easier to rear. They are also in high demand and are very nutritious.

*Source: Alabama Extension*

Catfish is one of the most suitable fish for fish farming. They are mostly farmed in fresh water ponds. They have a high market demand and have very high health benefits hence their high demand. The most common types of catfish that are farmed include; channel catfish and the blue catfish.

The sharp tooth catfish is the one, which is used instead of the Tilapia. This variety of fish normally has fewer issues but many benefits. The spawn production of the catfish is quite higher than that of the tilapia. Per spawn, the tilapia produces 200-1000 while the catfish produces 100000 eggs per month. This means that a small hatchery can produce many catfish. The growth of this kind is exceptionally fast and has a unique ability. This type of fish has the ability to breath in air and thus makes it possible for them to be transported in minimal water. These kinds of fish are not difficult to grow and can be easily raised by a farmer as hobby or for commercial purposes (Subasinghe, 2017). A catfish farmer can decide to grow them alone or rear them with other kinds of fish such as the trout, the bass and the bream. Catfish can be tended on either a half-time or a full-time basis. This normally depends on the size of operating that the farmer is dealing in. starting one's own fish farm of catfish can be as easy as modifying and stocking an existing fishpond. The first thing is that one needs to place a siphon in the pond in order to remove excess water in order to provide access to the bottom of the fishpond. The bottom needs to be very plowed and needs to be as flat as possible. In order to level up the field, one needs to fill up the deep ends and dig out the shallow ends. The other approach involves filling up the bottom with bentonite clay in order to prevent the water from being lost in the ground.

**Figure 1.4**: Preparation of a catfish pond. One should consider that the ground is a farm and that they have killed the native fish that was present in the pond.

*Source: Agricfarming.*

It is also important to kill out the native fish that was in the pond initially by adding a chemical, 3 pints of liquid rotenone for every foot-acre of water.

It should be then left for a week in order to make the process effective. After that, one can drain the water and then refill. After refilling the water in the pond, the farmer can then add the fingerling catfish. The stock of the catfish in the pond normally depends on how they are being fed. If the number is quite high then the farmer needs to increase the amount of food and the number of times to feed them. Another aspect, one should choose the schedule of the time for feeding the fish and a suitable place since the catfish tend to possess memories. In order to concentrate the fish, one needs to remove a third to a half of the water from the pond. It is ideal for the farmer to walk around the pond with a net in order to catch the fish.

Marketing of the catfish is a very important aspect given the fact that they are highly demanded. There are a number of ways of farming catfish, which include raising them in tanks, underwater cages and other vessels. However, the most ideal way of raising a catfish is by the use farms that are large enough to hold several acres of catfish ponds given the fact that they multiply in great numbers. There are more than 2000 species of catfish with more than 40 being reared in the United States. For the farmers that need to have a large-scale operation, it is ideal for them to acquire a large piece of land that is enough to hold suitable catfish ponds. Large surface areas are always preferred to avoid overcrowding and allow for breeding of the same. If possible, one can use the already available ponds or they can also dig up their own depending on what they find to be ideal. The supply of water should be adequate but should not reach the brim since the fish can get out of the pond and when they get to a place where there is no water, they can last only for 72 hours before they die. As such, it is quite important to always monitor them so that if there is a case in which there are some that get out of the water, the farmer can be able to return them as soon as possible (Stentiford et al., 2020).

In the case where one is using a natural source, the farmer needs to ensure that they screen out any fish that may try to enter the pond. When rearing catfish, it is very important to always aerate and heat the pond. One can be able to use a number of equipment such as the pump, propeller and paddlewheel aerators, which are known to stir the water and circulate the required dissolved oxygen to allow the catfish to breathe. The water should always be kept at temperatures between 82 to 86 °F in order to ensure that the catfish have optimal growth. The ponds that are set aside for spawning should be kept at lower temperatures than the other pond, which are between 77 to 81 °F. To save yourself from the process of aerating the ponds, one needs to keep smaller numbers of catfish. It is also important to feed the fish every day and the food pellets should contain all the necessary minerals and vitamins for their health. During the first few weeks of growth, protein is important for the fingerlings to boost their growth. It is also very important to add spawning equipment to your pond, which may be ceramic pipes, metal cans or plastic barrels but should be large enough to allow for the male and the female catfish to go inside and breed. In some areas such as India, the government was able to ban the catfish as the villagers complained that they pose a threat to other species especially the indigenous ones.

The salmon is popular especially the Atlantic salmon. The chinook and the coho are species of salmon that are also farmed. It is mostly farmed by individuals or companies in the northern part of America. There have been changes in many industries including the fish and it is also changing. The increase in population is seen to threaten food security in the world. Apart from that, the climate changes that are occurring are also seen to threaten the food systems. Salmon is the name given to a number of fish species that are under the family of Salmonidae. Salmon can be found in the wild and in the farmed sources. The two common varieties of salmon include the Atlantic salmon and the Pacific salmon but the Atlantic one is the most commercially available given the fact that they are commonly farmed. Its farming started in the 1960s as an experiment. The farming of salmon has become very common in the fishing industry such that, currently more than 60% of salmon is currently being produced in the world.

**Figure 1.5**: The salmon is highly demanded in most homes and restaurants. It can be bought as fillets or one can decide to buy the whole fish.

*Source: Freepik.*

A salmon farming production lasts for about 3 years. In the first year of production, the process takes place in the freshwater environments and after that, they are then transported to the seawater cages given the fact that they are naturally known to do well in salt water environments. At the time they reach harvestable sizes, the salmons are normally taken to the industries for processing in preparation for their sale. For the consumers, most of the salmon are normally sold in terms of salmon fillets although one has an alternative of buying a whole fish (Rimmer et al., 2013). The salmon is known to contribute towards food nutrition and security as it provides nutrient-rich foods that meet the global dietary requirements. They are known to provide high-quality and eco-efficient protein. They help meet the growing global demand for food products with the kind of nutrients it possesses. Consequently, the salmon provides continuous innovations that support healthier diets and more resilient global food systems. It is a food source that is celebrated in most countries as a delicacy.

Salmon farming is however considered to be bad for animal welfare and also considered to be very damaging to the environment. Waste from the salmon farms can be very damaging to the quality of the water. It is very risky given the fact that salmon are normally reared in large water bodies to increase the production of the same. Most firms also use salmons in the production of fish oils and cosmetics, which may prove very damaging to the environment. To be able to meet the future demands of salmon, it is required that production increases though it must be able to match the reductions in environmental impacts and improvements in resource efficiency. The methods that are chosen to rear the salmon need to be ones that can be able to support nutritious diets and provide sustainable food systems.

**Figure 1.6**: Salmon is very popular especially the Atlantic one. The farms are in crisis given the reports that they are costing the economies, the societies and the environment. The salmon can be very damaging to the environment.

*Source: Seafood Source.*

The tuna is a saltwater fish and the different types include; the yellowfin, albacore and the bluefin. The bluefin tuna is one of the most expensive fish on earth. Tuna fish farming requires a lot of skill and expertise as it is very complex. This is because simulating their natural environment is quite hard.

Tuna fish farming is normally done for the purpose of creating tasty and nutritious cuisines. The tuna fish is also in high demand given the fact that most individuals use the same instead of red fish. The tuna fish is a saltwater fish and basically is commonly known to belong to the Thunnini tribe. This tribe of fish contains more than 15 species of fish. The smallest type of tuna is known as the bullet tuna which weighs up to 2kgs and the largest is the Atlantic bluefin which weighs up to 700kgs which makes them very expensive. The Atlantic bluefin is the largest of the remaining species of tuna and shockingly this fish and is able to live for more than 40 years. If not protected, this species may end up extinct.

Just like the mackerel sharks, tuna fish have the capability of maintaining their own temperatures. This means that their temperatures can be quite higher than the surrounding water body. This fish has a streamlined sleeky body and is even listed as the fastest swimming fish. The yellow fin tuna is the one considered to be the fastest. The tuna fish is considered as an ideal replacement for red meat given the fact that it possesses the same nutrients but is much healthier than red meat. This fish is suitable for a number of health benefits including the fact that it is excellent for heart health as it aids in reducing the cholesterol levels in the body. It also aids in weight loss and relieves obesity problems, helps in good blood regulation, boosts immune system to increase the levels of energy in the body and prevents some types of cancers such as breast cancer. Other benefits of the tuna fish include the fact that it acts as the best home remedy to reduce stress and depression, prevents kidney diseases and arthritis and provides relief from inflammation problems (Rahman, 2014).

Because of the increased benefits of tuna fish, the demand for the same is quite high. Tuna farming can be done in every corner of the world but it is always beneficial to do the same in a place where there is a high market for the same. As such, it is always advisable to set up a farm where there is a good local market since it may prove beneficial to sell locally than to sell internationally. Since this fish is a salt-water fish, there should be access to salt water around the place the farming is to take place.

**Figure 1.7**: The blue fin tuna. A common type of tuna is commonly raised and is very expensive.

*Source: On the water.*

The eel is a carnivorous type of fish and it lives in freshwater or in the mouth of oceans. They are also known to be catadromous, which means they migrate from fresh water and into the sea to mate. The farming of eels is very common in Asia. China, Taiwan and Japan are some of the leading producers. The eels are known to only breed once in their lifetime. When they are still young and underdeveloped, they are glasslike. When this small, the ocean currents tend to transport them to the shores. These young ones are known as elvers and need to migrate to the fresh water lakes and rivers so that they can be able to feed and grow for some years. When farming these types of fish, the farmer buys the underdeveloped eels where he isolates them to see whether they have been infested by any kind of disease or pest. In the case where this is not done, if the disease is passed on the already established eels, the farmer may undergo high losses. They are considered to be ideal fish to be farmed given the fact that they are tolerant to many conditions including being cramped in large numbers. Growing eels in ponds may be ideal given the fact that the warm temperatures required by the eels can be maintained.

There is an increased demand for fish around the globe. This has led to the increase of fish farms but as well as to satisfy demand there is overfishing in the oceans as well as lakes and some individuals have even set up wild fisheries. This then leads to the decrease in the fish stocks especially in their natural habitat and this in turn will affect the aquatic habitat (Roubach et al., 2019). In fish farming, it involves setting up artificial fish colonies and the fish are provided with sufficient amounts of food, protection from predators such as snakes, eagles, kingfishers and many others, the fish also get veterinarian services in case some of them get sick or die and the fish colonies are easy to harvest. The fish that are raised in farms do not influence the numbers that in the wild hence they help in the conservation of fish and also help to decrease the demand by providing an alternative to the market.

Fish is produced generally worldwide but China is the largest producer of fish as it supplies more than 60% of the total global supply of fish. By the year 2016 in the global supply of fish, 50% of the fish supplied was produced by aquaculture. Some farms are known for the rearing of fish that are carnivorous such as the salmon and they are usually fed fishmeal and fish oil that is known to be extracted from wild fish.

There are two known categories of fish farming which are local production, which depends on the food that are is internally that is within the farm while the other category is intensive that basically depends on the food that is produced externally. In intensive fish farming, the farmer can increase their production as long as there is sufficient water supply and enough oxygen. In this instance, the farmer will be required to have a water purification system, which will be integrated, into the farm. Because fish need water with oxygen for them to survive and to grow in size the farmer, can sometimes bubble, cascade flow the water and sometimes add aqueous oxygen.

There are parasites that affect fish such as fish lice, intestinal worms, bacteria and protozoa. The farmer needs to then ensure the water supply is purified and treated to avoid such parasites. In this instance, it means that the farmer has to be an expert or in some cases hire a manager that is competent and the farm has to have daily and intensive monitoring. Due to all this and the recycling of water, it means the operational cost of a farm will be very expensive.

Some farmers have opted to raise coldwater ornamental fish such as the stripped fish, goldfish also referred to as koi and the multicolored fish. In this instance, venturing into such a business looks very lucrative as it produces higher income per weight of the fish produced. The most common countries that have koi farms are Israel, the United Kingdom and Germany.

The cage system is also referred to as offshore cultivation if the cages are placed in the sea or ocean. In this instance, the fish cages are placed in ponds, rivers, lakes, oceans and bayous so that they can contain and protect the fish. The fish are kept in cages until they are harvested. The fish are fed, monitored up until they reach the market size where they are then harvested.

In recent years, cage, farming in the seas and oceans has been gaining popularity (Rodrigues et al., 2012).

Although there has been acceptance of this method there has been concern raised especially when the fish that was caged escape in dams or in open water. A good example is when approximately 300,000 non-native salmon escaped their cages in Puget Sound in Washington State. It is then believed that the non-native salmon fish that escaped would risk endangering the native pacific salmon species because the non-native salmon will reproduce faster gain weight a lot and tamper with the genetic composition of the native Pacific salmon species.

Recently there have been improvements in the type of cages that have been used but due to the unpredictability of weather and other factors, the risk of damage to the cages and escape of the fish is always present.

The use of copper-alloy nets, in recent years copper alloy nets have been the most important netting material in pisciculture. They are mostly used because they are antimicrobial which means they can destroy algae, viruses, fungi, bacteria and other microorganisms. The nets also prevent organisms from growing on them and this means the environment is cleaner and healthier as compared to the traditional nets which required labor-intensive regular cleaning. They also contain other properties such as being corrosion resistant properties and they are generally very strong in marine environments.

In aquaculture, farms in countries such as the USA and countries in South America and Asia deploy the use of copper-zinc brass alloys. There are different copper alloys that are being researched on if they can be used. Copper-nickel and copper-silicon are still in research with trials and assays being conducted on the two.

The use of pond systems or irrigation ditch, these include fishponds that are used to raise fish. This involves having a ditch or a pond that is able to retain water with a constant supply of ground water. In this type, after a ditch or pond has been dug it is then lined with bentonite clay or some farmers decide to place plastic or PVC sheets at the bottom and sides of the pond to prevent water from escaping. The fish are often fed commercial food if it is large scale or they may decide to grow algae and plankton to serve as fish food. To small-scale farmers, they may use kitchen waste and sometimes they even incorporate vegetables.

**Figure 1.8**: Fishponds. These are ponds used for rearing fish.

Fish are known to be sensitive to the type of water that is supplied to the pond. The water quality should always be monitored i.e. to check the fertility, clarity and pH of the water. The fish will grow substantially especially if the water does not undergo eutrophication or the oxygen levels do not go low (Rosa et al., 2012).

Composite fish culture, this a technology that was developed in India in the 1970s by the Indian Council of Agricultural Research. In this system, it involves both local and imported fish, where up to five to six different species of fish are kept in the same pond. The fish that are selected do not come from the same habitat and hence they do not feed on the same type of food, which means the fish in the pond do not compete for food. Examples of fish that are commonly used in this system include; common carp who feed at the bottom of the pond, rohu which normally feed in the column, the silver carp which normally feed on the surface as well as catla among other different species. Some of the fish species even feed on the excreta of other fish such as fecal matter from the common carp.

Classic fry farming, it is also known as the flow-through system and it involves trout and other sport fish and they are raised from eggs to fry or fingerlings, they are then carried by trucks to streams where they are then released. The fries are usually raised in long and shallow concrete tanks, which are normally fed by clean stream water.

In some cases, the fish can be raised indoors. The ponds or large fish tanks are kept inside large structures. In these structures, the water is often treated and recycled. The treatments include ozonation, ultraviolet treatment and oxygen injection. This means the water has to be recirculated but the recirculated water has to undergo periodic exchange of the water. These large fish farms tend to have aquaponics, which means that plants are grown using liquid instead of soil. They also use denitrification. These two methods when used will help to reduce the amount of nitrates in the water.

**Figure 1.9**: Indoor fish rearing where the temperature is regulated and it is covered on all sides.

After the fish have grown to full they are then harvested and there are different methods that are used during the slaughter of fish. The methods that are used for slaughtering have been divided into two where one is considered to be inhumane and the other humane. The inhumane methods based on the animal rights activists include; air asphyxiation which involves keeping the fish in the open air this makes the fish suffocate. The fish can take up to 15 minutes to die although they become unconscious quite fast. Some individuals harvest the fish and keep them in ice baths or put them in water that is near freezing point. Most of these individuals explain that the cold water prevents the fish from decaying fast. Exsanguination involves the fish being removed from the water and as they are held they are cut to allow for bleeding. All these methods are considered inhumane as fish are seen to suffer before dying (Rico et al.,2017).

The more humane methods include; percussive stunning which is rendering the fish unconscious by hitting it on the head. Electric stunning is divided into two i.e. dry stunning which involves passing a current of electricity in the fish for a period of time and wet stunning which involves passing the current in the water while the fish are still inside. The application of current while the fish is in water requires more current as compared to dry stunning.

## 1.3 CRUSTACEAN FARMING

This involves the farming of crustaceans i.e. crabs, lobsters, krill, woodlice, barnacles, shrimps, prawns and crayfish. They belong to the kingdom animal, phylum Arthropoda, clade Pancrustacea and subphylum Crustacea. They are mostly free-living aquatic animals but some of them are terrestrial e.g. woodlice. Crustacean farming production is in both marine and in fresh water habitats where marine produces approximately 70% and fresh produces 30%. The vast majority of the crustaceans which is almost 80% of the crustaceans produced are made up of only two species i.e. the giant tiger prawn and the white leg shrimp.

**Figure 1.10**: A prawn. A type of crustacean reared for consumption.

*Source: Animals Australia*

In the world production of crustaceans, almost 70% comes from Asia mostly from China and Thailand. Crustacean farming has been going on for several centuries, especially in Asia. In most instances, the prawns and crabs have been caught as juveniles and raised in captivity until they are fully mature. The "tambaks" which are very common in Indonesia are a very well-known traditional practice of raising crustaceans. In the western countries at the turn of the century, there have been advances in technology especially in marine biology and this led to the increase in the number of marine fisheries which then enabled the development of production of larval stages of the American and European lobsters and it was produced in large scale (Raja & Jithendran, 2015).

There are several steps that need to be considered for the crustaceans that are usually farmed: 1) obtain the seed stock or commonly known as breed stock, 2) maturing, spawning, hatching and the rearing of larvae,3) past-larval rearing or the nursery phase then 40 the grow-out culture and lastly 5) harvesting and processing.

Before starting a crustacean farm, an individual has to have a breed stock. A good example is in Ecuador where farmers catch the shrimp using nets. This is very lucrative as the wild stock is far more expensive as compared to hatchery stock mainly because they have a higher survival rate. After they have been captured, they are kept in hatcheries where the females are allowed to lay eggs. After they have laid the eggs and the eggs hatched there needs to be security and predictability. The farmer has to be very skillful in order to understand the different species. In some, the eggs hatch in a few hours and they law quite a large number of eggs whereas the caridean shrimp lay few eggs for and they bear them for a number of weeks and a few months until they hatch.

In the two groups, some species mature, mate and eventually spawn while they have been kept in captivity but in some, they need to have a controlled environment for them to mature, mate and spawn.

**Figure 1.11**: A lobster. A common delicacy in Chinese restaurants. Also raised in farms.

*Source: The University of Marine*

Apart from shrimp, another common crustacean is the lobster. In lobster farming, the farmer has to locate where to set up the farm. This location has to have space, clean water that is free from pollutants and there has to be a source close by where the farmer can get lobster larvae. This then makes most of the farms to be close to the ocean because of the proximity to lobster larvae and the lobsters like living in salty water. Before establishing a farm, some countries require farmers to have permits to raise the lobsters. After being granted permission then the farmer can set up the troughs he/she will use, where the troughs should be solid so as to keep the lobsters inside and any predator from accessing the lobsters (Quesada et al., 2013). The troughs should be arranged in such a way that it is easily accessible and then should be filled with clean salt water. If the farm is close to the ocean the farmer can raise the lobster in ocean pens. The farmer should use seawater because of its salinity because water from other areas has to be salinized. The pH of the water should be maintained between 6.5 and 8.5. There should be weekly maintenance of the troughs so as to keep it clean. The water should be continuously filtered so as to circulate oxygen and remove the waste. The farmer can then purchase larvae from the local supplier. The lobsters after hatching have a tiny, transparent and spiny body with large eyes. It will then undergo four different stages before it becomes or looks like a lobster. They should be fed once a day. A netting should be placed at the top of the trough to protect the lobster.

Another type of crustacean that is farmed is the crab. The most common is the mud crab. The crab is tasty and it is consumed in many countries and hence it is very lucrative as it can earn a lot of foreign exchange.it has many advantages including labor costs are very low, the production cost is very low and the crabs grow very fast. There are various types of these crabs i.e. green mud crabs and the red claw, which are very common.

**Figure 1.12**: Cages that are used for raising crabs. They are floating on seawater and arranged in water for easy feeding.

*Source: The daily star*

It is simple to establish a mud crab farm is relatively cheap and easy to establish. They are raised based on the facilities that are readily available in the area. There are two systems that are commonly used; grow out system, in this system the crabs are raised and grown for a period of 5 to 6 months when they have reached the market size where they are harvested. They are kept in ponds and a fence is kept around the pond to provide security and prevent predators and unwanted guests. The feed used is low-cost fish, small-sized crabs and shrimps too.

The other system is the fattening system; this is the raising of soft-shelled crabs until their shells become hard. The hard-shelled crabs have more market value compared to the soft-shelled crabs. It takes less time to raise them. The crabs can be fattened in the pond where the crabs are raised on the shore where the water is between 1 and 1.5 meters high. A fence is placed to prevent the crabs from escaping and they are stocked with young soft shell crabs from the nearby market. The second is fattening in pens and cages. They can be raised in floating cages. The cages are arranged in rows so that they can be easy for feeding (Perumal et al., 2015).

## 1.4 MOLLUSK FARMING

This basically involves shellfish farming. The species that are normally cultivated include Ostreidae (oysters), Mytilidae (mussels), Pectinidae (scallops), and Veneridae (clams).

**Figure 1.13**: Oysters are becoming very nutritious to the world population. They are easy to grow and mature very fast. Farmed oysters are preferred given the fact that they are cleaned and are free from contamination.

*Source: Food republic.*

Oyster farming is done around the world but close to the oceans and seas. In today's world, there appears to be a notion that farm-raised seafood is not as good as wild-stuck seafood. The media and plenty of hobby businesses have raised issues for aquaculture, commonly mentioning its environmental effect and dangers to purchaser health, however, this generalization is not actual for all varieties of seafood aquaculture, particularly now no longer oysters. Oyster aquaculture is extraordinarily one of a kind from finfish aquaculture. Farmed oysters are a completely sustainable desire and similarly adorable in comparison to their wild counterparts. The fun fact about the cultivation of oysters is that they are normally cultivated in their natural environment. In addition to this, there is no need to provide feeds to these organisms. This is due to the fact that they eat the same kind of food as the normal oysters that include algae. The farming of this kind of fish is normally different from the finned fish given the fact that they are not put in cages. The only thing that is done is that the oyster farmers put them in oyster cages that prevent predators from harming them. These cages are also appropriate as it protects them from being smothered by silt to the bottom of the ocean. Most farmers of the same use the bottom-plant technique, which enables the oysters to grow on the bottom of the ocean just like the natural oysters.

The farming of oysters has been known to reduce the pressure that is put on wild fisheries. It is unlucky that a lot of our wild oyster fisheries were overfished or destroyed, however, now there are policies and a motion to repair them back. With the boom of oyster aquaculture, farmed oysters have substantially decreased the want to depend on wild oysters. This has allowed time for wild oysters to get better and repopulate beds that had been as soon as decimated a few years ago. This will enable the oyster population in marine waters to increase (Pandey et al., 2014). In addition to this, oysters are known to provide benefits to the surrounding ecosystem. Most of the critics of aquaculture point out that, there are many negative impacts of aquiculture on the environment. They argue that farming of the same will harm the environment. Oysters however control phytoplankton growth because they have the ability to fill up a gallon of water per hour. It helps in controlling the algal growth, which may affect the quality of water. The farmed oysters benefit because they are normally fed well, cleaned and protected making them live for longer periods than the other oysters.

Mussels are known to be easy to farm and are very good to eat. This kind of shellfish is known to grow in clusters and have a black shell. Mussels are also known to be able to clean the water as they are filter feeders. These fish species are known to collect tiny microorganisms from the water. Mussels are almost similar to oysters but they are known to never leaving their natural habitats. To keep away from having to dive deep into the sea to head and get them, mussel farmers use floating rafts from which lengthy ropes hold deep into the sea. Newly spawned toddler mussels float in with the current, searching out a gap to settle down – and because it turns, out, big, cumbersome ropes make for the appropriate home. Soon enough, they all have a gap on their new putting skyscraper. They by no means even contact the sea floor – this means that no dredging is required. The location of the rafts is very important given the fact that the organic matter that the mussels feed on flows with the current. Their nutrition mainly comes from plankton. The mussels attain full growth once they are 8 months old. The farmed mussels are normally bred high above the ocean floor, which makes them free from grit, sand and other impurities. This makes their flavor pure and they have a plump texture. As such, most of the consumers of the same are normally seen to prefer the farmed mussels to the natural ones.

Scallops are the other types of shellfish that can be bred. Their cultivation is a commercial activity and are known to be very expensive. Traditionally, fishing for wild scallops has been the desired practice, because farming may be expensive (Panigrahi& Azad, 2007). However, international declines in wild scallop populations have resulted in the boom of aquaculture. In China, the bay scallops are primarily farmed. This is considered as one of the best types of aquiculture given the fact that there is no damage to the bottom habitat of the fish. Most of the sea scallops are harvested by way of dredging the ocean floor and is not considered to be very suitable. This is given the fact that it can result in considerable by-catch of the non-targeted species and sadly damage the bottom of the ocean. Calico scallops are currently under a huge risk of going extinct given the fact that they are highly farmed.

Clam farming is a common type of mollusk farming where it starts from the hatchery where the larvae are set on fine screens for the down-dwellers. The young seeds are then raised in up wells, which allow the algae-rich water to be pushed above through the juvenile clams. This allows these clams to feed in a continuous manner and hence grow rapidly. Once the clams have been able to grow, they are then transferred to the mesh bottom trays where they are taken to the nursery rafts in the bay. The moment the clams have been removed from the nursery rafts, they are then suspended from the rafts, which then allow them to grow without having them to raising food from the hatchery. Once the clams develop to fingernail length, they are shipped to the developing areas - in this case, William Bay, Washington. Here they're planted via way of means of sprinkling them over net panels on an incoming tide in order that the infant clams then work their way through the mesh of the net into the sand wherein they're then shielded from predators consisting of crabs and geese as they develop.

## 1.5 ALGA CULTURE

Alga culture is a form of farming that involves cultivating algae species. There are different species of algae such as those that fall under the microalgae and the macro algae categories. The microalgae involve the phytoplankton, microphytes or also known as plank tonic algae. Macro algae on the other hand entail the cultivation of seaweed. The process of cultivating and harvesting seaweed is also known as kelp farming. Some farmers practice the farming of algae in a simple form by cultivating batches found naturally from the ecosystem while some are more advanced such that they have involved themselves in producing and managing the full life cycle of algae species.

Seaweed farming is mostly carried out for commercial purposes, as it is associated with many industrial uses. In addition, sea weed farming is used for other purposes as in food production as it comprises essential minerals and nutrients such as omega 3 fatty acids or natural food and colorants. For other purposes, the grown crop is used for creating dyes, fertilizers, bio plastics, chemical feedstock, and pharmaceuticals and algal fuel. In controlling pollution the algae could also be used. The majorly cultivated types of seaweed taxa include *Eucheuma* spp., *Kappaphycu salvarezii* spp., *Gracilaria* spp., *Saccharina japonica*, *Undaria pinnatifida*, *Pyropis* spp. and *Sargassum fusiforme*. Mainly most of them are produced for food apart from *Graciliaria* which is cultivated for agar purposes and *Eucheuma* along with *K. alvarezii* for making a gelling agent also known as carrageenan. China, Indonesia and the Philippines have been the top leading countries in the production of seaweed worldwide (Ottinger et al., 2016). Other countries that have produced seaweed notably include South Korea, North Korea, Japan, Malaysia and Zanzibar as well as Tanzania. The practice of algae farming has been continually embraced and developed as an alternative farming system in improving the economic conditions of farmers as well as in reducing the economic pressures associated with fish farming. Fisheries have been overexploited in the past and thus the embrace of seaweed farming.

Generally, due to the ongoing demand, the most cultivate algae category is the seaweed with an output of about 30 million tons in 2016. Due to the growing importance in the quality of algae culture, most farmers prefer to do monoculture production. The process of obtaining the species to plant is gotten from first investing in mixed cultures then the species that becomes dominant with time is they are chosen by the farmer to be the species to be mainly produced. In the event that a non-dominant species would be believed to have a particular value or use, the pure cultures of the species are also cultivated. For research purposes the most used in making a conclusion about the algae species are the individual species that are obtained through the process of serial dilution. The procedure is that first, the personnel involved with the process dilutes either a wild sample or a lab sample that contains the desired algae with filtered water then, small aliquots are introduced into a large number of small growing containers. After dilution, the microscopic examination procedure follows. This happens from the source culture that is used to predict that a few of the growing containers may contain a single cell of the desired proposed species to be worked on. There after the cultivators use a microscope after exposing the contents to a light table so as to identify the containers to be used in the larger production processes.

**Figure 1.14**: Farming of algae by use of agricultural technology.

*Source: Future of agriculture*

Alternatively, in identifying the species to culture, farmers may use a special medium that operates by excluding other various organisms including the invasive algae. For example, the species of *Dunaliella* genus is widely grown and has high adaptive levels to places with alkaline or salty conditions that few organisms grow. Otherwise, as for larval mollusks, mixed algae culture can be the best to be used. First, the cultivator filters the seawater to remove algae that appear very large for the larvae to consume. Secondly, the cultivator adds nutrients while it also aerates the results. After two days or several days, the mixed algae would have generated a thin soup containing ready to use larvae for algae production purposes. The method is convenient as it is associated with low costs that include low maintenance cost purposes.

Among the microalgae, brine shrimp species are the most cultivated as they produce many eggs in the larvae production stage. The eggs are then hatched in accordance with the demand in the market or they can also be feed to cultured fish larvae and crustaceans. Although different algae have different uses in farming them as well as the requirements involved, water, carbon dioxide, light as well as carbon dioxide are important factors that cut across all algae production. The basic formula that occurs in basic reactions in algae production in the water entails carbon dioxide + light energy + water = glucose + oxygen + water. The process of growing these algae is called autotrophic growth. However, some species of algae that consume sugars, for example, glucose can be grown without the light factor put into considerations. The practice of farming the various species without using light is known as heterotrophic growth (Opiyo et al., 2018). However, the water used in the production should be in a 15°C and 35°C temperature range for the production of algae to go on effectively. In a typical algal-cultivation system, such as an open pond, light only penetrates the top 3 to 4 inches (76–102 mm) of the water, though this depends on the algae density. This is because the growth and multiplication of the algae in the water may block light from reaching deeper into the waters as the species become dense. For most algae, direct sunlight is too strong, as past studies have reported that algae only use a tenth of the energy for its growth, which is coming from direct sunlight. Nevertheless, they have also concluded that exposing the algae to direct sunlight during production is the best method in ensuring a strong growth of the algae as the algae growing underneath the surface is able at least able to reach more light created from the shade of the other algae species growing above.

Further scientists have recommended the use of deeper ponds so that grows would agitate the water, circulating the algae so that it does not remain on the surface of the water. Basically, the paddle wheels used in the water settings stairs the water and compress the air from the bottom thus lifts algae from the lower regions. This helps the algae in the lower regions to grow well as also prevents the algae in the above from overexposure to the direct sunlight if the practice is carried out continuously. Instead, glowing plates that are made from sheets of plastic or glass can be placed within the tank in order to offer precise control over the light intensity and also distribute light evenly to the growing algae species. However, this kind of lighting system is rarely used in cultivation due to its associated high costs. The bad odor that is generated from bogs, swamps and stagnant waters is mainly caused by oxygen depletion due to the decay of deceased algal blooms. In anoxic conditions algae culture which contains inhibiting bacteria breaks down the organic contents thus producing hydrogen sulfide and ammonia materials, which results in the odor (Otta& Hoque, 2020).

The hypoxia majorly results in the death of aquatic species. Researchers have found that in places where algae are intentionally cultivated, maintained and managed for expected output the eutrophication or the hypoxia rarely occurs. Anabaena species also known as the blue-green algae with odorous chemicals sometimes produced from living algae and bacteria. The majorly found causing chemicals are the MIB also known as *2-methylisoborneol* and *geosmin*. These two give a musty odor that can be very strong sometimes. Additionally, in the event that the cyanobacteria die, it would release additional gas that is trapped in the cells. The chemicals are detectable in the lower levels and are responsible for the issues in drinking water treatment, distribution as well as safety in the health, taste and odor. These bacteria may produce toxins that have been a problem in drinking water.

Nutrient requirements in the growth of algae species are such as nitrogen, phosphorus and potassium. The nutrients applied in the form of fertilizer boost the growth of algae. Other traces of elements such as silica and iron are also essential in the growth of aquatic life. This is because deficiency of the elements may inhibit the growth and productivity of a given location where the cultivation is ongoing. Carbon dioxide availability in great amounts helps in the growth of algae species at a fast rate. Besides, the elements must be able to dissolve into the water being used for cultivation in bioavailable forms for algae to be produced optimally.

The IPCC special report on the ocean and cryosphere in the changing climate has recommended further attention on seaweed farming as it is a carbon-negative crop that has a high potential for climate change alleviation. The attention as a mitigation tactic would be imperative. The elements have an easy access to the raceway-type ponds and the lakes. The open ponds are greatly susceptible to contamination by other microorganisms, such as other algal species as well as contamination from bacteria to which they are exposed. Famers thus opt to carry out their activities in the closed systems for the growth of monoculture processes. This enables them to be able to be in control over the temperature and lighting of the species under their cultivation. The output preseason is dependent on the site thus in the tropical areas growth of algae is limited to the months that are warmer.

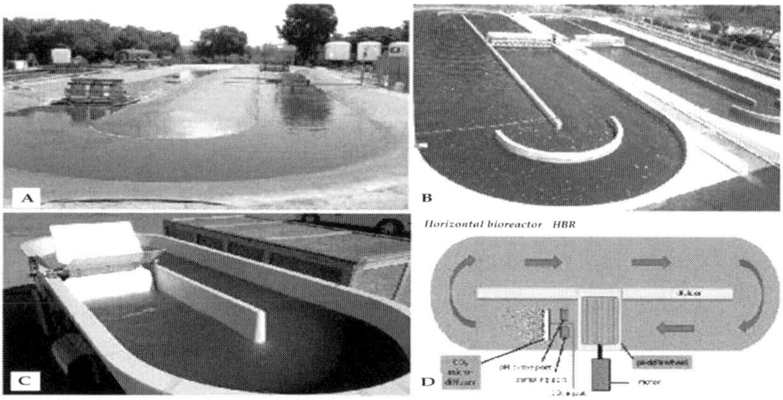


**Figure 1.15**: Algae farming by use of ponds. They are quite cheaper to construct and generates the most output if the ponds are big.

*Source: ResearchGate.*

Although open ponds are cheaper to construct, requiring a trench or a pond basically, the use of large ponds generates the most output as compared to other systems requiring the same kind of cost of production. Scientists in the past have found out that the use of open ponds in cultivating could exploit unusual conditions that could suit the growth of the specific algae. For example, the *Dunaliellasalina* spp. which grows in extremely salty water conditions, in the open pond the other types of organisms are likely to be excluded allowing only the growth of the specific species. The system of open farming in the ponds is also recommended if the cultivator has a way of harvesting only the desired algae while leaving the rest. This could also happen if the ponds are frequently re-inoculated before invasive organisms could be produced to a large extent. The latter approach is from time to time applied by chlorella cultivators in the growing conditions for the chlorella, which is not exclusive of the competing algae.

In harvesting, some chain diatoms can be filtered though as the stream of water is flowing through an outflow pipe. In the efforts of allowing other algae to escape pillowcase of a fine mesh cloth is tied over the outflow pipe. The chain diatoms are then held in a bag and feed shrimp larvae and inoculate new tanks and or ponds. When a pond is enclosed with a transparent or a translucent covering the area would then operate as a greenhouse for rearing algae. The conversion allows for mitigation of challenges associated with ponds that are open systems. This allows for greater output of the species to be generated and gain dominance in the production space at the pond water settings. The growing species may also exceed the growing season, such that if heated the pond could produce year-round. A photo bioreactor may be used in growing algae because it may be used to incorporate light sources. This involves all the virtually translucent containers used in growing algae. The term is mostly used to refer to the closed containers as opposed to the open ones that have tanks or ponds. They provide all the required nutrients including the CO. This makes it possible for the cultivators to operate their farming systems in batch forms such that they can be able to grow and harvest continuously as the harvest sets in (Ngajilo & Jeebhay, 2019).

The operation in the continuous harvesting and control of the farming requires careful deliberations of all elements so that operations in the farming may remain effective. The cultivator must ensure that correct rates of elements are placed into considerations including sterilized water, nutrients, air, carbon dioxide and lighting. Given all these requirements, the regulator can then operate for long periods continually producing algae optimally. Researchers have concluded that algae that grow in the log phase or longer time periods contain greater nutrient contents than the old algae that renew their growth after a short period of time. Maximum productivity at the farm level occurs when the exchange rate which is the time taken to exchange one volume of liquid is equal to the doubling time which can be interpreted as the in mass or volume or the algae output (Kalbassi et al., 2013).

The use of PBRs in algae culturing provides an environment for which the suspension can provide substrate culture from a bio film. The bio films that are generating as a result of the PBRs are likely to produce higher yields as compared to other means of production systems. Bio film-based PBRs have the advantage that they can produce far higher yields for a given water volume, but they can suffer from problems with cells separating from the substrate due to the water flow required to transport gases and nutrients to the culture. The different known types of suspended culture PBRs include tanks, polyethylene sleeves or bags and Glass or plastic tubes. Bio film PBRs contain packed bed as well as porous substrate PBRs. Filled bed PBRs can be different shapes, including flat plate or tubular. In Porous Substrate Bioreactors (PSBRs), the bio film is exposed directly to the air and receives its water and nutrients by capillary action through the substrate itself. This avoids problems with cells becoming suspended because there is no water flow across the bio film surface. The culture could become spoiled by airborne organisms, but shielding against other organisms is one of the functions of a bio film.

Algae are harvested using micro screens this is done by means of centrifugation then by flocculation and by froth flotation. The milled shells of crustaceans are processed to acquire chitin, a polysaccharide found in the shells, from which chitosan is derived via de-acetylation. The water in the pond which is more brackish or saline requires larger amounts of flocculants. Flocculation is often too expensive for large operations. In flotation, the cultivator aerates the water into froth, and then skims the algae from the top. The algae farming processes are thus economically profitable (Narbonne et al., 2021).

Chapter 2

# FISH STOCKING

## CONTENTS


2.1 Introduction .......... 30

2.2 Scale of Protection and Assistance in Mariculture .......... 31

2.3 Fish Stocking .......... 31

2.4 Production Steps in Fish Stocking .......... 34

2.5 History .......... 39

2.6 Stocking Today .......... 41

2.7 Overfishing .......... 42

2.8 Benefits if Fish Stocking .......... 44

2.9 Sea/Ocean Ranching .......... 50

2.10 Valliculture .......... 53

2.11 Floodplain Fisheries Management .......... 56

2.12 Brush Parks or Acadjas .......... 57

## 2.1 INTRODUCTION

In most bodies of water, fish are abundant. Cusk-eels and snail fish can be seen in practically all aquatic settings, from high mountain streams (e.g., char and gudgeon) to the abyssal and even hadal depths of the deepest oceans (e.g., cusk-eels and sailfish), however, no species has ever been discovered in the deepest 25% of the ocean. Fish have a higher variety than any other class of vertebrates with 34,300 species identified.

Fish, notably as food, are a vital resource for people throughout the world. Commercial fishers and livelihood fishers forage in the wild or grow fish in ponds or in enclosures in the ocean (in aquaculture). Recreational anglers catch them, and they are maintained as pets, nurtured by fish keepers, and displayed in public aquariums. Fish have played a significant part in culture throughout history, functioning as deities, religious symbols, and topics of art, literature, and film.

Tetrapods evolved from lobe-finned fishes; hence they are cladistically fish as well. Fish, on the other hand, are generally made paraphyletic by removing tetrapods (i.e., the amphibians, reptiles, birds and mammals which all descended from within the same ancestry). Because the term "fish" is characterized negatively as a paraphyletic group in this way, systematic biology does not regard it as a legitimate taxonomic grouping unless it is used in the cladistic sense, which includes tetrapods. The traditional name Pisces (sometimes ichthyes) is a typological categorization, not a phylogenetic one.

Ocean ranching is a sort of fish farming in which young fish are released into the ocean unprotected and unsupported in order to mature and be caught later.

Mariculture in pens or other contained and protected habitats is explicitly excluded from this description. We'll refer to this form of operation as simply ocean fish farming in this paper, and the unsupported, unprotected variety of ocean fish farming as ocean ranching. Both, on the other hand, fall under the umbrella of ocean fish farming.

It should be highlighted that protection and assistance are, at least in theory, continuous variables that can vary from zero (no security and aid) to extremely high degrees of protection and assistance (representing extensive protection and assistance). Ocean ranching is near the bottom of the scale, while ocean fish farming is near the top (Mohapatra et al., 2013).

In the middle, there exist variations that are difficult to categorize as either ocean fish farming or ocean ranching. Consider a group of fish that has been released into an open bay and is sometimes fed at specific locations. As a result, the fish choose to stay around rather than move away, despite the fact that they get much of their nutrition from the local habitat.

## 2.2 SCALE OF PROTECTION AND ASSISTANCE IN MARICULTURE

We use enclosure as the line of demarcation between ocean ranching and ocean farming in our report. This implies that the operation is classified as ocean ranching if the fish are not artificially contained.

According to the above definition, ocean ranching appears to be divided into two types, each with distinct economic implications. The first is when the fish are meant for the releasing agency's own harvest (or other usages). This is proper ocean ranching, which is quite comparable to land-based cattle and other animal ranching. This is referred to as "pure ocean ranching" in this research. The second sort of ocean ranching is when the fish are intended to be gathered by people who aren't affiliated with the agency that released them (Mishra et al., 2017). This practice is basically stock augmentation, thus this study will refer to it as stock augmentation or stock enhancement ocean ranching. It is critical to understand that the ideas of pure and stock augmentation ocean ranching, as well as ocean ranching and ocean fish farming, are essentially opposed poles on a continuum of options, with certain examples impossible to categorize as one or the other.

The economics of ocean ranching is the subject of this report. It is essentially focused on the first type of ocean ranching, which is pure ocean ranching. However, most of what is said also relates to stock augmentation ocean ranching, which is covered in detail when looking at other countries' ocean ranching experiences.

## 2.3 FISH STOCKING

Fish stocking is the process of breeding fish at a hatchery and then releasing them into a river, lake, or ocean to supplement current populations or establish new ones. Stocking can be done for commercial, recreational, or tribal fishing, but it can also be done to restore or raise the population of threatened or endangered fish in a water body that is off-limits to fishing (Michael et al., 2014).

Government entities may supply public waters with fish, whereas private groups may stock private waters with fish. Fish stocking in public waters generates a common-pool resource that is competitive but non-excludable. This means that on public grounds, everybody can feel and benefit from the advantages of fishing so long as fish continue to be replenished.

**Figure 2.1**: Stocking fish in a river in California.

*https://en.wikipedia.org/wiki/Fish_stocking#/media/File:BTC_Stocking_Site_in_California_off_Moabi_Park_exit_about_one_mile_from_the_border_on_I40_(2).JPG*

A fish hatchery is a facility for artificially breeding, hatching, and rearing animals in their formative years, particularly finfish and shellfish. Hatcheries raise larval and juvenile fish, shellfish, and crustaceans for the aquaculture sector, where they are moved to on-growing systems like fish farms to achieve harvest size. Pacific oysters, shrimp, Indian prawns, salmon, tilapia, and scallops are some of the common species bred in hatcheries.

Global aquaculture production was valued at US$98.4 billion in 2008, with China taking the majority of the market share; nevertheless, the value of aquaculture hatchery and nursery production has yet to be determined. Additional hatchery output for small-scale domestic usage or conservation programs, which is particularly prevalent in Southeast Asia, has yet to be assessed (Marques et al., 2020).

There is a lot of interest in replacing overexploited fish stocks by releasing juveniles that were either caught in the wild and raised in nurseries before being transplanted or created entirely in a hatchery. In the United States, finfish larvae culture has been used successfully in stock enhancement attempts to replace native populations. To aid in the conservation of native fish species, the US Fish and Wildlife Service has established a National Fish Hatchery System. Hatcheries raise larval and juvenile fish and shellfish for transfer to aquaculture facilities, where they are matured to harvestable size. The industry gains from hatchery production in three ways:

**Figure 2.2**: Tanks in a shrimp hatchery.

*https://en.wikipedia.org/wiki/Fish_hatchery*

*Out of season production*: a reliable supply of fish via aquaculture plants is critical to the market. Broodstock conditioning can help farmers get more juveniles by extending the natural spawning season. Supply can also be ensured by sourcing from hatcheries in the opposite hemisphere, with opposing seasons.

*Genetic improvement*: some hatcheries use genetic modification to raise the quality and productivity of farmed species. Artificial fertilization aids in the development of selective breeding programs aimed at improving production traits such as growth rate, disease resistance, survival, color, higher fecundity, and/or a younger maturity age. Selective breeding, hybridization, and other genetic modification techniques can all help to improve genetics (Lulijwa et al., 2020).

*Reduce dependence on wild-caught juveniles*: aquaculture accounted for 46% of total food fish supply in 2008, or over 115 million tonnes. Although wild captured juveniles are still used in aquaculture, concerns about the sustainability of collecting juveniles, as well as the unpredictable timing and size of natural spawning episodes, make hatchery production an appealing alternative to meet the industry's expanding demands.

## 2.4 PRODUCTION STEPS IN FISH STOCKING

### 2.4.1 Broodstock

The technique of putting adults into spawning conditions by encouraging the development of gonads is known as broodstock conditioning. Broodstock conditioning can also be used to extend spawning beyond typical spawning periods, or to produce species that have been bred outside of their natural geographic range in diverse environments. Some hatcheries gather wild adults and bring them in for training, while others keep a breeding stock on hand. Broodstock is conditioned by keeping them in flow-through tanks with ideal light, temperature, salinity, flow rate, and food availability (optimal levels are species-specific). Another key part of broodstock conditioning is guaranteeing the production of high-quality eggs to promote larvae growth and survival by optimizing broodstock health and welfare. The nutritional status of the mother is frequently a determining factor in egg quality. To boost larval survival rates, high quantities of lipid stores are required.

Broodstock management entails tinkering with the environment around the broodstock in order to maximize survival, improve gonadal development, and boost fertility. This type of conditioning is required to ensure the long-term viability of aquaculture production, and to raise the quantity and quality of eggs produced and control the timing of maturation and spawning.

One of the most important steps for aquaculture to assure its growth is the management of technology for gamete production in captivity. Unfortunately, most fish that are raised in captivity have some form of reproduction malfunction. In aquaculture, many species of captive fish can achieve reproductive maturity, and gonadal growth happens regularly (Luis et al., 2019). However, some female species frequently fail to reach the last stage of oocyte maturation. Because of the economics of broodstock management, hormone modification and acceleration of ultimate oocyte maturation are significant. In Salmoniformes, for example, the requirement to retrieve eggs by stripping is a significant constraint, and the timing of ovulation must be precisely predicted, as over-ripening can occur minutes or hours after ovulation. Controlling broodstock reproduction is therefore critical for the long-term viability of commercial aquaculture.

The biology of the species must be considered while selecting species to use. This covers their mature size, breeding mechanism, feeding habits, and ability to withstand harsh situations. Farms must also decide whether to raise their own broodstock or purchase it from wild populations. When natural populations are excluded, the farm can be regarded as a self-contained unit free of genetic influence from outside sources.

Pond-reared broodstocks are chosen as immature juveniles and nurtured to sexual maturity in acceptable conditions. These organisms require consistent water conditions as well as a well-balanced, species-specific, protein-rich feed. As juveniles develop, this improves the germinal tissue for subsequent seed stock.

The pond or tank where the broodfish are kept must be large enough to accommodate and condition the broodstock. You may need to adjust the number of individuals and, in certain cases, segregate the sexes, depending on the species. Broodstock males and females might be treated to varied conditions depending on their gender. Male and female sturgeons, for example, react to hormone levels differently, allowing for more command over eggs and sperm.

It is necessary to modify the parameters of the water in which the mature broodstocks are kept. The aquaculturist must take into account the proper oxygen saturation, temperature, and pH of the water, all of which might vary depending on the species (Liao et al., 2009).

It is necessary to modify the parameters of the water in which the mature broodstocks are kept. The aquaculturist must take into account the proper oxygen saturation, temperature, and pH of the water, all of which might vary depending on the species. To postpone spawning, artificial light can be utilized to vary the apparent day duration and convey distinct seasonal aspects. The temperature of the water can be raised for the same purpose. Female broodfish are frequently stressed and underweight after spawning. To ensure survival until the next spawning season, they demand extra care and nutrition at this time.

Managers can choose reproductive qualities that influence an individual's ability to produce eggs and improve fecundity by providing them with the right environment and food. This is also conceivable in pond-reared species, where qualities such as fecundity can be chosen across generations.

The breeding season and spawning timings can be altered, extending the producing season. Because the market is available all year, this results in more efficient aquaculture. Hormonal therapies can help to speed up the spawning process by two to three weeks. Manipulation of the photoperiod can change spawning time by up to four months and is inexpensive and simple to do.

To generate more desirable fish, broodstock managers might use or opt for qualities such as quick growth rates or disease resistance over generations. This ability to improve stock genetics is more efficient and results in higher-value stock. Broodstocks also allow you to plan and control all matings selectively. Selective breeding is a crucial aspect of aquaculture species domestication (Li et al., 2011).

Predation is reduced in pond-reared broodstocks, which can be a substantial cause of death in wild populations. They also gain from the elimination of variable environmental impacts. Keeping broodstock in an easily accessible pond or tank ensures that breeding adults are available anytime they are needed.

Broodstocks are subjected to different selective pressures than normal when used to replenish wild populations. As a result, they may not be fit enough to live in the wild, or they may modify and reduce natural genetic diversity owing to the cut-off nature of breeding from a smaller population.

Broodstocks require outside supplementing on a frequent basis to avoid the detrimental impacts of closed populations. Broodstock domestication in hatcheries can impair reproductive ability and change other genetic features. For example, a trout stock that had been kept as a closed population for 20 generations produced fewer and smaller eggs (Labh & Shakya, 2014).

## 2.4.2 Spawning

Aquatic animals discharge or deposit eggs and sperm into water, which is known as spawn. To spawn is a verb that relates to the process of releasing eggs and sperm, and it can refer to both sexes. With the exception of aquatic mammals and reptiles, most aquatic creatures reproduce by spawning.

Many aquatic species produce spawn, which is made up of their reproductive cells (gametes), some of which will be fertilized and produce offspring. Females discharge ova (unfertilized eggs) into the water in vast quantities, while men release spermatozoa (milt) to fertilize the eggs concurrently or sequentially.

Most aquatic animals, including crustaceans like crabs and shrimps, mollusks like oysters and squid, eukaryotic organisms like sea urchins and sea cucumbers, amphibians like frogs and newts, aquatic insects like mayflies and mosquitoes, and corals, which are basically small aquatic animals rather than plants, procreate by spawning. Fungi, such as mushrooms, are said to "spawn" a white, fibrous substance that serves as the substrate for their growth. There are many different ways to spawn, based on sexual anatomical differences, how the sexes interact, where and how the spawn is discharged, and whether or not the spawn is protected.

Natural spawning can happen in hatcheries during the typical spawning season, but the spawning of mature animals can be produced using a number of methods when additional control over spawning time is desired.

The following are some of the most frequent methods:

1) ***Manual stripping***: in the case of shellfish, the gonads are usually removed and the gametes retrieved or rinsed free. The anesthetized fish can be physically stripped of eggs and sperm by massaging the anus under the pectoral fins, causing gametes to easily flow out (Akter et al., 2016).
2) ***Environmental manipulation***: thermal shock, which occurs when cool water is exchanged with warmer water in flow-through tanks, can cause spawning. Alternatively, if natural spawning cues are known, they can be simulated in the tank, for as by adjusting salinity to replicate migratory behavior. This method can cause a large number of individuals to spawn, but it also increases the risk of uncontrolled fertilization.
3) ***Chemical injection***: a variety of substances can be employed to stimulate spawning, the most frequent of which being hormones.

## 2.4.3 Fertilization

Fertilization, also known as generative fertilization, syngamy, and impregnation, is the union of gametes that results in the birth of a new individual organism or progeny and the beginning of its development. Before the union of gametes, processes like insemination or pollination are sometimes referred to as fertilization. Sexual reproduction is the process of fertilization and creation of new individuals. By the mechanism of vegetative fertilization, the haploid male gamete joins with two haploid polar nuclei to generate a triploid main endosperm nucleus in angiosperms during double fertilization.

Eggs can be gently cleaned before fertilization to remove wastes and germs that could taint cultures. To maintain genetic diversity in hatchery-produced stock, it is vital to encourage cross-fertilization among a large number of individuals. Separate batches of eggs are kept separate, fertilized with sperm acquired from multiple males, and left to stand for an hour or two before being examined under a lens to ensure the highest rates of fertilization and estimate the number of eggs to be transported to larval rearing tanks (Anderson et al., 2019).

### 2.4.4 Larvae

Rearing larvae through the early stages of life is done in nurseries, which are usually affiliated with hatcheries for fish culture, but shellfish nurseries can operate on their own. Nursery culture of larvae to rear juveniles of a size acceptable for transfer to on-growing operations can be done in a number of ways, including completely land-based systems or larvae that are later transferred to sea-based rearing systems that eliminate the requirement for feed. The survival of juveniles is reliant on very high-quality water. Feeding is a crucial step in the raising process. Although several species can survive on maternal reserves alone (lecithotrophy), most commercially available species require feeding in order to maximize survival, growth, productivity, and juvenile quality. Nutritional requirements differ depending on the species and larval stage. Live prey is regularly provided to carnivorous fish; because of their small size, rotifers are usually given to early larvae before moving on to larger Artemianauplii or zooplankton. Because it is a labor-intensive procedure, producing live feed on-site or buying-in is one of the highest costs for hatchery plants. Artificial feeds were developed to cut the costs of live feed manufacturing and improve nutritional consistency, but these alternatives have been found to have worse growth and survival rates (Amenyogbe et al., 2018).

### 2.4.5 Settlement of shellfish

Shellfish hatchery production also includes a critical settling phase, during which free-swimming larvae limb out of the water onto a platform and, if proper conditions are discovered, complete metamorphosis. After metamorphosis, the juveniles are referred to as spat, and it is during this time that they are sent to on-growing operations. A variety of factors influence settlement behavior, including substrate type, water velocity, temperature, and the availability of chemical cues that indicate the presence of adults or a food source, among others.

As a result, hatchery operations must be aware of these cues in order to induce settlement, as well as be able to substitute artificial substrates for easy handling and transit with little mortality.

**Figure 2.3**: Juvenile salmon towards the end of their stay in a hatchery.

*https://en.wikipedia.org/wiki/Fish_hatchery#/media/File:CapilanoRiverRegPark-salmonhatchery.jpg*

## 2.5 HISTORY

Trouts were originally supplied in wilderness lakes in the western United States in the 1800s. For nearly a century, this activity has been conducted out solely for the aim of growing and increasing sport fishing, with little consideration for the environmental implications. Following the dawn of a new environmental awareness in the 1960s, traditional fish-stocking procedures were challenged, first by federal land management departments, then by their state equivalents, based on evidence that revealed negative impacts on the biota directly related to introduced fishes (Assefa & Abunna, 2018). The pragmatic ethic that guided resource management till the 1960s gave way to one that recognizes the value of all life forms and their ecological complexity, a viewpoint that is now shared by many anglers. The need for wilderness fish stocking is now widely debated, particularly in light of shifting social values and objectives. If invasions continue to destroy the richness of mountain lake ecosystems, future generations will have fewer options.

Fish stocking has been used for hundreds of years. The site and number of fish introduced were not carefully documented during the first hundred years of stocking; the sole purpose of stocking was to improve sport fishing irrespective of ecological consequences such as biodiversity degradation. "When trout planting was originally introduced, the country was seized by a very utilitarian resource management mentality that put short-term human interests above nearly any other consideration," says biologist Edwin Pister. The US Fish and Wildlife Service, as well as state fisheries departments, have recently improved their record-keeping of which species of fish are stocked where.

This began in the 1960s when studies revealed that fish stocking had a negative influence on the ecological complexity of other living forms. The Wilderness Act of 1964 also made the public aware of the negative effects of stocking on other organisms. As a result, fish stocking is currently the subject of great controversy, as it comes with a variety of costs and advantages.

Howard Zahniser of The Wilderness Society drafted the Wilderness Act of 1964. It established the legal definition of wilderness in the United States and provided federal land protection for 9.1 million acres (37,000 $km^2$). This began in the 1960s, when studies revealed that fish stocking had a negative influence on the environmental complexity of other living forms. The Wilderness Act of 1964 also made the public aware of the negative effects of stocking on other organisms. As a result, fish stocking is currently the subject of great controversy, as it comes with a variety of costs and advantages.

Howard Zahniser of The Wilderness Society drafted the Wilderness Act of 1964. It established the legal definition of wilderness in the United States and provided federal land protection for 9.1 million acres (37,000 $km^2$).

The National Wilderness Preservation System was established when Congress enacted and President Lyndon B. Johnson signed the Wilderness Act on September 3, 1964. The Act classified 9.1 million acres (37,000 $km^2$) of national forest wilderness areas in the United States of America that had previously been protected by administrative orders as statutory wilderness areas. Currently, the NWPS has designated 757 places as wilderness, including 109.5 million acres of state-owned territory in 44 states and Puerto Rico (5 percent of the land in the United States).

Storage of non-native fish for sport and food in the U.S. was only launched in 1871 by the establishment of the United States Fish Commission. Spencer Fullerton Baird, Head of the new agency, was tasked with "researching and suggesting remediation strategies for the reduction of food fish in the Sea and Lakes." On Cape Cod, Massachusetts, Baird erected his Woods Hole headquarters. There he carried on research of striped bass, blue fish and many other commercial and sports fish with his team of scientists and researchers. They have developed a 255-page report on the fish resources of the United States. The Congress awarded the $15,000 in food supplies to the team, which was later successfully introduced into United States lakes and rivers by non-native species such as rainbow trout, salmon, streaked bass, and carp. Sport clubs and private individuals supplied fish in the early years (Balcázar& Ma, 2007).

The distributor of fish today is the responsibility of state fish and wildlife authorities together with hatcheries. And their aim was to plant as much fish as feasible into as many water bodies as possible until recently. Now that the invertebrate and amphibian populations are aware of the adverse impacts of the store of fish, it is carried out far more selectively.

## 2.6 STOCKING TODAY

The introduction of non-indigenous species is now much more thoughtful because it might severely damage the fragile indigenous population; approaches tend to be more sustainable. Storage is used to restore indigenous species to waters that were overfished or could no longer be cultivated. The storage procedures "give and take" are those in which fish are only stocked and then re-alimented. In response, most countries have implemented policies prohibiting fish stockpiling in regions that could impair aquatic life or ecosystem diversity and promote stockpiling in aquatic bodies where there is no danger. For example, Trout Unlimited has a policy of stating that "when a body of scientific data indicates that stocks of native biodiversity in historically non-salmonid streams should stop." Although many organizations remain exclusively focused on offering good possibilities for fishing, policies and attitudes move towards the integrity and conservation of resources.

## 2.7 OVERFISHING

Overfishing involves the removal from the water of a species at a speed that the species cannot replace, which results in the under population of those species. FAO estimates that a third of the world's fish reserves were overfished (FAO, 2018). State funds amount to around EUR 30 billion each year for fishing.

**Figure 2.4**: 400 tons of jack mackerel caught by a Chilean purse seiner. Overfishing catches too many fish simultaneously, so that the breeding population is too weary to recover. Overfishing is typically combined with inefficient types of commercial fishing, which are subsequently thrown away into enormous quantities of undesirable fish or other animals.

*Source: https://en.wikipedia.org/wiki/Overfishing*

Excess fishing can take place in water bodies of all dimensions such as lakes, rivers, lakes and oceans. This can lead to resource depletion, lower biological levels and low rates of biomass. Persistent overfishing may lead to significant dependencies, which are no longer sustainable for the fish population. Some forms of overfishing, such as hawk overfishing, have caused the disruption of whole marine ecosystems.

It is dependent on whether ecosystem parameters can recuperate from overfishing. Fisheries can recover from overfishing. Dramatic changes in the composition of species could lead to an ecosystem transition where other energy balance flows involve mixtures of species different from those that were there before the original stock reduction. For instance, after trout are overfished, carp may take over so that the trout cannot restore the breeding stock (Kalbassi et al., 2013).

Overfishing has removed a lot of stocks from the fishing industry worldwide. In a 2018 assessment, the Organization for Food and Agriculture projected that 33.1% of the world's fish inventories were over-fished. In pre-industrial times, significant overfishing was seen. The overfishing of the Western Atlantic Ocean in particular has been widely documented since the first days of European colonization in America.

There was a decreasing tendency of a fraction of fish that are in bio-sustainable levels, from 90% in 1974 to 66.9% in 2015. In comparison, the share of fished stocks at bio-unsuitable levels grew to 33.1% in 2015 from 10% in 1974 with the greatest growth in the late 1970s and 1980s.

In 2015, fished stocks (previously referred to as fully-fished stocks) accounted for a maximum of 59.9% and fished stocks represented seven% of the total stocks assessed. Whereas the fraction of fished stocks steadily declined between 1974 and 2015, the fish stocks plummeted from 1974 to 1989 to a maximum level of sustainable fish, up to 59.9% in 2015.

The largest proportion (62.2%) of inviolable stocks was found in the Mediterranean and the Black Sea in 2015, followed by 61.5% in the South-East Pacific, and 58.8% in the South-western Atlantic in 2015. On the contrary, there was the least share (13 to 17%) of organically unsustainable fish stocks in Eastern Pacific, Near-East Pacific (Area 67), North-West Pacific (Area 61), Western Central Pacific and South-western Pacific.

The fisheries expert Daniel Pauly, who was known for pioneering human impacts study on global fishing, said: it's like we deploy our military for fighting creatures in the water. We win this fight gradually to destroy them. And, for nothing – for no reason – this is a bit disappointing to witness this destruction occur. Curiously enough, all these reversible effects, all the species that have disappeared are coming back, all the small animals will grow, all the relations you can see will remerge and the system will emerge again.

An introduced species, alien species, exotic species, adventive ones, immigrant species, foreign species, non-indigenous or non-native species is a species that has been produced by human activity, whether directly or indirectly, whether inadvertently or unintentionally, outside of its native distribution range. The effects of non-native species on the local ecology may be different.

The introduced species which are established and expand beyond the initial area are regarded as "naturalised." The technique of introduction to humans is different from biological colonization in which organisms spread through 'natural' (non-human') methods such as tempests and rafting into new locations. The term "neobiota" in Latin expresses the property that the biological linkages (e.g. food web) are new to their area in relation of these species. The neozoa (also neozoons, sing. neozoon, e.g. animals) and the neophyt can be classified further into neozoans (plants) (Jana et al., 2018).

There is a highly varying effect of imported species. Some have a significant negative influence on a local ecosystem (in which case species are more properly classed as an invasive species), whereas other species introduced can have no negative effect or small effects (no invasiveness). Certain species were purposely introduced to fight pests. They are termed bio controls and might be considered as an alternative to pesticides, for instance in agriculture. In some cases, there is no long-term potential to be beneficial or harmful. Scientists, governments and others have examined the consequences of invasive species on natural habitats.

## 2.8 BENEFITS OF FISH STOCKING

Fish stocking offers the society a range of benefits. Fishing is a common sport for 101.6 million American people over the age of 16 partaking in fishing and hunting activities. In addition, an 8% rise in angling involvement between 2011 and 2016 was noted in a U.S. Department of the Interior report. In 2017, little over 49 million Americans, 11.9 million of them young, took part in some fishing. The majority of fishers' fish around four and 11 times a year, with others reporting more than 100 trips (Hossain et al., 2020). Fishing, in particular for young fisherman, is more entertaining by adding fish to public lakes and streams. In addition, stocking can restore endangered, threatened, or native fish. The red list of endangered species of the Union for Conservation of Nature contains 1,414 fish species at risk of extinction. The storage of these into lakes, rivers and streams can support currently threatened populations and reduce the number of species at risk. Many of the fish that are typically utilized for storage also have poor reproduction rates and are overfished if not stored each year. In addition, since stock fish include larger trophies, more fishermen are more prepared to pay for a fishing authorization, which means state fisheries agencies have more money to expend on the management of natural resources. In 2018, there were roughly 30 million paid license holders in the US, grossing $720 million that year.

**Figure 2.5**: Fishing for stocked trout in Wyoming.

*Source: https://en.wikipedia.org/wiki/Fish_stocking*

Many fish species, such grass carp and catfish, assist clean up water through the use of algae and other green creatures. Algae can take stagnant swamps, attract insects and unpleasantly look at lakes, rivers and swamps. Many people opt to store some fish species in order to treat them. For people who appreciate a range of water activities, this provides a favorable externality. The economic viability of fish stocking has also been studied by studies. In Stockholm University it is concluded that the simple economic analysis suggests that Volga pikeperch stock can be profitable, with an annual interest rate of 43%. These authors also note that greater stock fish populations reduce labour and equipment expenses related to each fishing activity. They discover furthermore that stocking can re-establish food web interactions specifically for pikeperch at a more "natural" level, where herring numbers are reduced and zooplankton is flourishing, hence benefiting the whole ecosystem.

**Figure 2.6**: Fish are stocked all over the world. This river in Mount Aspiring National Park, New Zealand, is stocked with trout.

*Source: https://en.wikipedia.org/wiki/Fish_stocking*

Fish stocking has further advantages. Fishing equipment including rods, bobbins and lures are spent in millions of Dollars across the country through licensing charges (to the profit of state governments). The International Game Fish Association and Bass Anglers Sportsman Society are members of fishing societies who also like fishing in stronger bodies of water. In spite of the foregoing facts, Virginia scientists have found that 80 percent of the fish in its stockpiled rivers are still native, even with the rainbow trout stock. The Minnesota Ministry for Natural Resources found out, as organic rules are applied to choose which and where to stock, that stored musky can cohabit. Until recently there was little understanding of the ecology of lakes and rivers (Hemaiswarya et al., 2011).

Fish must be stored wisely, in locations where they can be enjoyed but do not present a threat to natural species, in order to maximize the benefits of storage and cost control.

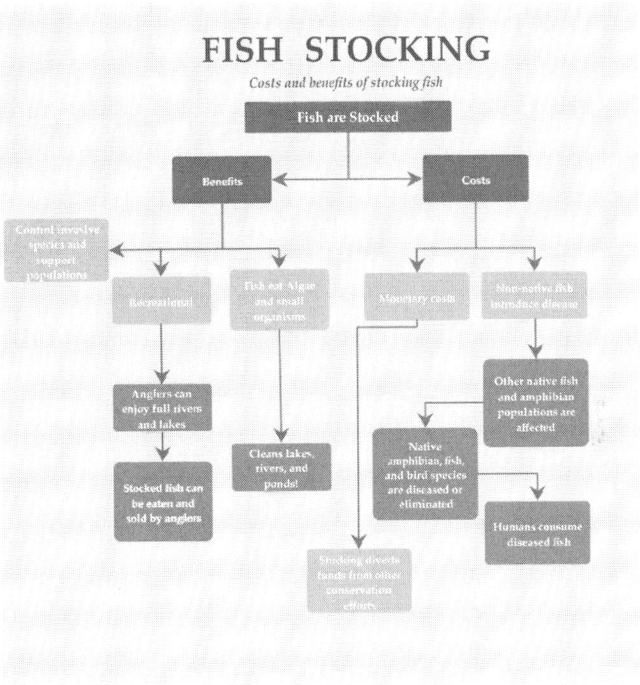


**Figure 2.7**: Costs and benefits of fish stocking.

*Source:https://upload.wikimedia.org/wikipedia/commons/f/fc/Costs_and_Benefits_of_Stocking_Fish.jpg*

The PA Fish & Boat Commission alone in Pennsylvania is planned to store 4,398,227 trout into its streams and lakes in 2019 (brooks, browns and rainbows). Lake Ontario is expected to get 2 767 660 salmon and trout, one of the five large lakes, in 2019 as well. State agencies have estimated that approximately 1.7 billion fish were stored in the United States in 2004. A total of 43.65 million pounds of fish was released, mainly in the western Member States, with 104 different species of fish stored. Tree species, bass, salmon, muscular salmon, walleye and numerous species of panfish are prevalent in the United States today stored for recreation.

Fish are produced in a fish hatchery before they are distributed throughout the country in rivers, lakes, swamps and sometimes seas. Just as people have a demand for food, so do fishermen. When supplies are low in a certain body of water, fish are transferred from hatcheries to their corresponding sites in a huge water tank or aircraft. Costs connected with storage are typically handled by fishermen's associations, commercial fishers' associations,

State fishing and wildlife agencies, and sometimes government subsidies (Hasimuna et al., 2020).

Around 32 percent of the world stock of fish is over-exploited, depleted or recovers, and has to be urgently reconstructed, according to the United Nations Food and Agriculture Organization. Some people hail fish farming as a solution to the problem of overfishing. But farms far from benign can damage ecosystems seriously by bringing illnesses, pollution and alien species. Depending upon the fish species, how it is harvested and fed, the amount of production and the location of the farm, the damages produced by fish farms vary.

A major issue is that farms typically depend on lower-than-food wild fish species, such anchoves, for feeding the bigger carnivorous farmed species rather than reducing the impact on wild populations. One pound of fish like salmon or bass can be grown by up to five pounds of smaller fish. The overfishing of these smaller "drilling fish" has implications for the ocean's entire environment.

As in the case of industrial livestock farms on land, fish are typically contained with limited space to move in an excessively crowded and tight environment. Injuries, fine damage and other vulnerable injuries may be experienced. Farmers handle pesticides and antibiotics through overcrowded and stressful conditions, leading to disease and parasite outbreaks, such as sea lice. The usage of antibiotics can produce drug-resistant strains of illnesses that affect populations of wildlife and even human beings who eat farmed fish.

Fleeing fish brings an additional environmental concern. Hundreds of thousands of fish escape every year from farms and endanger indigenous species' genetic diversity and survival. High storage densities lead to substantial contamination caused by fish discharges and unused food, resulting in bad quality water with high ammonia and low oxygen content. Outdoor fish farms may also attract marine predatory creatures like seagulls and marine lions that are sometimes poisoned or shot by fish-eaters.

Evidently, the belief that fish experience no pain remains popular, despite evidence to the contrary. Methods of slaughter are horrible in the aquacultural field. There is little attention to the agony of animals and most of them are completely aware of the several minutes of murder. Some animals, such as salmon in the US, also starve to empty their intestines for many days before they are slaughtered. Fish are not usually stunned and are bloody, repeatedly struck, suffocated or freezingly die by bleeding. There are no standards to ensure the humane treatment of fish in the United States as in many other countries (Freitas et al., 2020).

The existence of this highly overfished species is endangered by years of the unregulated and underreported catch of bluefin tuna in the Mediterranean and Atlantic oceans. Ranching bluefin tuna is a popular industry to supply the increased demand for sushi in Japan and internationally, worsening the situation. Fishermen use longlines and bags to catch the them before they reach breeding age. They are then housed in seascapes for 3–6 months, and they are fattened before they are killed and transported by thousands of pounds of smaller wild fish.

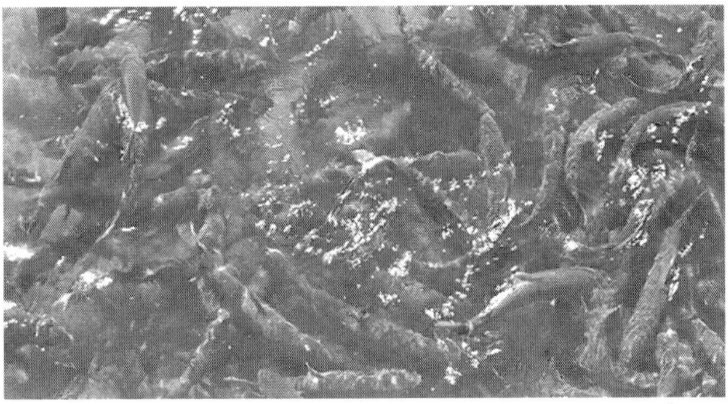

**Figure 2.8**: Fish farming is a form of aquaculture in which fish are raised in enclosures to be sold as food. It is the fastest-growing area of animal food production. Today, about half the fish consumed globally are raised in these artificial environments. Commonly farmed species include salmon, tuna, cod, trout and halibut. These "aquafarms" can take the form of mesh cages submerged in natural bodies of water, or concrete enclosures on land.

*Source: https://awionline.org/content/fish-farming*

Unlike aquaculture involving the cultivation of aquatic life within controlling environments or commercial production of certain aquatic species by strictly controlling the majority of their life histories, traditional fisheries enhance production in natural environments by controlling and transplanting or releasing a part of the life history of certain species. The juvenile fish generated in hatcheries are permitted to proliferate and grow in natural food until it reaches a harvestable size in fresh, brackish, or marine environments. In contrast to aquaculture, culture fishing is not confined to the pressure of land or population and does not require changes to or management of the system of culture in order to resemble the natural environment as nature itself is used. In addition, in contrast to aquaculture, harvests are variable and the returns are harder to estimate as "the release of fish into the sea results in a reduction of 1-15%, depending on species and locale, of fish being harvested."

Even so, traditional fisheries are more used to improve the fisheries resource environment and replenish natural stocks whose populations have fallen as a result of overfishing or ecological destruction, or simply to maximize water body productivity, whether it be an open bay, a coastal lagoon or a reservoir for freshwater. In fact, its implementation or extension was designated as a high priority on the development agenda in a number of locations worldwide throughout the future years.

The most important types of cultivable fisheries are:

- maritime/ocean farming as exemplified by the valliculture of the Mediterranean;
- storage and restoration of freshwater seas and reservoirs and
- floodland fishing.
- marine/ocean fishing as exemplified by valliculture in the Mediterranean.

## 2.9 SEA/OCEAN RANCHING

Marine ranching, which frees juvenile fish from the seabed to produce in marine waters unprotected on natural foods, is used as an important aquatic pasture by the sea. While practiced in different ways around the world, juvenile fish are freed up to the sea through ponds, clubs or channels in their distinctive form and are permitted to grow until they are ready for marketing (Faruk et al., 2004).

### *Salmon*

Salmon culture is common in nations like Norway, Chile, Alaska, Washington and Oregon, Canada, Japan, the Soviet Union, New Zealand, Iceland and the Baltic States. As a result of its successful artificial spawning combined with the rapid development and high level of the home precision it is able to maturity harvesting, the Pacific salmon (*Oncorhynchus* sp.) is the most common kind utilized for sea-farming.

In the northern countries where salmon smolts were discharged into the Baltic Sea area, salmon ranching took place over 100 years ago. Hatchery salmon production in the United States stretches back to 1872 when its first hatchery on the McCloud River in California was created by the American Fish Commission. Following this in the 1877 Rogue River, Oregon, the first private salmon hatchery. Japan was introduced in 1877 and Western Canada in 1884 with technology for the artificial spread of salmon.

The outcomes were not considered a success in Japan while the hatching and release of salmon have been performed for almost 100 years. The only way to make major increases in salmon output was through the innovative approach of intermediate feeding and timely release of salmon fries from 1961. Most Japanese hatcheries induct larval salmon (alevins) and employ gravel-lined bunches to feed freshly emerged chum fry on artificial diets (Deb et al., 2021). The use of technologies similar to Japan is developed in Soviet hatcheries in equal numbers. Chinook and coho salmon in the Pacific Northwest are the major species utilized in hatchers, with prominent raceways and swamps feeding young fish. Floating race paths must have been created for Alaska in protected saltwater bays.

Due to successful hatchery operations, the salmon farming business of Canada has increased considerably during the past five years (Annett, 1989). The creation and deployment of spawning and Egg incubation channels have been pioneered by Canadian hatchery workers who give the eggs and alevins with favorable environmental conditions.

The spawning channels generate offspring with fast growth and good survival capabilities because they replicate high-quality natural spawning bedsIncubation canals, shallow races with gravel beds without silt, allow alevins from artificially fertilized and hatched ovens to be developed in a natural substrate. Before releases into marine water (e.g. sockeye, coho, and chinook) the salmon fry that needs to be grown should be moved from the incubation to feed-in systems made of tanks, traces, ponds, floating styles or combined or modified with them. This should be done in low-salinity or fresh water. The young fry is frequently fed high-protein diets that contain vitamin supplements and specific lipids.

## *Other Species*

Other marine species were employed for ocean ranch in Japan, except from the famed Pacific salmon. In 1953, the Japanese National Fishery Cultural Project succeeded in creating mass seed production techniques and later release into the Seto Inland of young fish of commercially valuable seafoods including the Kuruma creeps, red sea bream, blue crabs, soles, flower and yellowtail. The project presently reaches the stage of the release of brood and control the spreading activities by the fishers themselves. Based on its past results, the project has been nationwide expanded through the building of other prefectural fishing centers throughout Japan (Dauda, 2020).

The *Haliotis discus* (Ezo abalone) is the largest ocean breed species in Japan, released in 1980 in rocky sections of the Japanese Sea surrounding Hokkaido around 1.15 million hatcheries. The gigantic Ezo Scallop (*Patinopecten yessoensis*) is likewise spread over the Okhotsk Sea, Nemuro channels, and elsewhere on the Hokkaido shore.

## *Coastal Lagoon Farming*

Extensive lagoon cultivation, with an annual production of about 10 000 tons, is a traditional vocation in different European and Mediterranean countries. Italy has the most important brackish water areas (called "valli") amongst the Mediterranean countries, which are used as a lagoon for the cultivation of commercially valuable species such as *Anguilla* spp., *Mugilcephalus*, *Liza aurata*, *Liza saliens*, *Liza ramada*, *Chelonlabrosus*, *Dicentrarchus labrax*, and *Sparusaurata*.

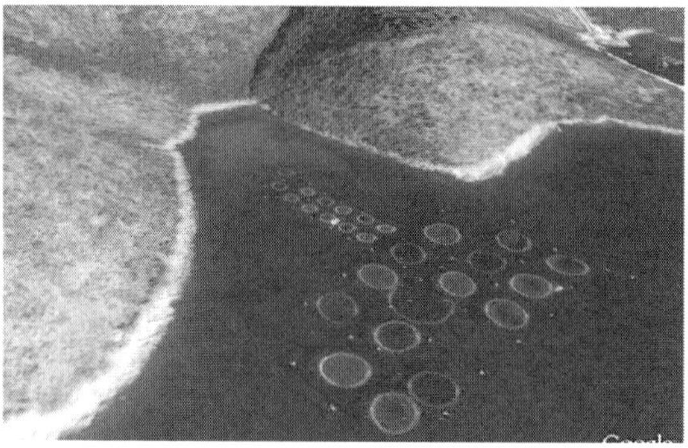

**Figure 2.9**: In view of wild-caught seafood overfishing globally, ocean fishing has increased fresh fish to western markets and become one of the world's fastest-growing sectors. Precise reporting of wild caught fish quantity was hard and we asked whether in the statistics for farmed fish production comparable errors exist in data. There is widespread marine fish farming in the Mediterranean and stationary cages from the coast of 16 nations may be observed using Google Earth satellite images. Example image from Google Earth showing a fish farm off the coast of Greece.

*Source: https://www.researchgate.net/publication/221844785_Fish_Farms_at_Sea_The_Ground_Truth_from_Google_Earth*

The relocation and artificial recovery of valley/coastal lagoons take place by anadrome, natural and yearly movement. A valli catches mostly depend on the amount of fingerling that entered the lagoons some years previously and on the amount of fingerling stored in them.

## 2.10 VALLICULTURE

Valliculture is one of the oldest forms of aquaculture in the Mediterranean region, as the practices are named in view of their existence as a fish culture. Its origins extend back to the first rudimentary fish tubs in Adriatic and Tyrrhenian coastlines and fattening systems. The upper Adriatic populations were produced to utilize seasonal migration from the sea to the lagoon and delta areas more conducive for their growth. These techniques were used to develop the techniques. The fish returned to the water due to an altered environment (temperature) or reproductive circumstances of the sea (Dauda et al., 2019). To use these regular migrations, extensive sections of brackish waters were surrounded to ensure that fish would not return to the sea and complicated permanent catch systems, fish barriers were built which constitute barriers to adult catching through channels communicating with the sea. Subsequently, a man-made seeding for friezes, fished elsewhere, was brought into the basins for several years from the basic fry ponding, that went freely into the lagoon from the sea. Successful management by hatchery fry with major species of coastal lagoons is believed to carry significant global promise for future expansion.

### *Brush-park fisheries*

Brush park fishing is a traditional method of low-tech aquaculture practiced in various parts of the world in coastal lagoons and in brackish waters. Brush parks in a variety of shapes and sizes are made. However, in principle, a brush park consists of an inner core or concentrated circles, densely packed tree branches or other debris, encircled by an external, more significant timber frame, regularly fished by encirclement.

The most complex varieties of brush-park fishing are found in coastal lagoons in Benin. In Benin there are several fundamental varieties of brush park fishing, which are together known as 'acadjas.' In structure, design and fishing characteristics they differ from one another. The acadjas are built of tree springs with tougher woods that create the periphery and soft wood springs with many ramifications that constitute the interior of the acadja.

In shallow tranquil waters with a depth of no more than 1.5 m, acadjas are preferred.

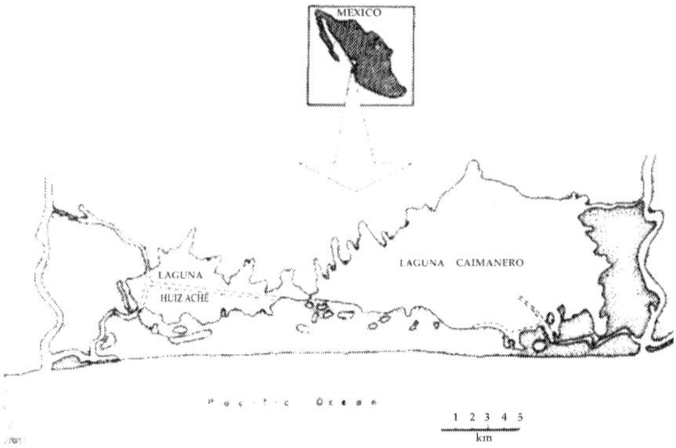


**Figure 2.10**: In the lagoon, the engineering work also serves a range of goals such as providing greater fishing surfaces and the passage by the neighboring lagoons of fisheries and shellfish. Interconnecting waterways also provide better access to nursery regions for larval fish and shrimps. Sedimentation reduction and circulation enhancement are other aims of internal canals. Hydraulic engineering within lagoon systems is exemplified by the Huizache-Caimanero lagoon complex in Mexico.

*Source: http://www.fao.org/3/X6855E/X6855E03.htm*

Fishing is carried out by enclosing the smaller types of acadjas with a net that then removes all branches and pursues the net. For the bigger brush parks the surrounding net is gradually caught by shifting the net towards the inside as branches are removing, and the fish are concentrated in small regions. The branches utilized are replaced and new branches are added every fishing cycle (Bossier &Ekasari, 2017).

The coastal lagoons of Nigeria, Côte d'Ivoire, Ghana, Togo and Madagascar are also used for brush parking fishing elsewhere in Africa. Brush-park-fishing is found elsewhere in Sri Lanka in the Lagoon of Negombo and was imported to several lagoons in Mexico.

## *Stocking of Inland Waters*

The stock and recovery strategy has been one of the most commonly utilized and supported practices for the management of lake and reservoir fishery resources worldwide.

The high yield of herbivorous, detritivores and plankton feeders (such as tilapia and carp) are often stored in lakes and reservoirs in the emerging countries of the tropics. This is in stark opposition to advanced countries of temperate areas where carnivores such as trite and bass are stored for recreational or sports fishing in lakes and reservoirs. In ASEAN countries, for example, large lakes and reservoirs were filled with hatchery-produced fingerlings, including tilapia, different Chinese-Indian carps, milkfish, mullet and catfish. (ASEAN Countries: It is an intergovernmental organization of ten Southeast Asian countries: Brunei, Cambodia, Indonesia, Laos, Malaysia, Myanmar, the Philippines, Singapore, Thailand, and Vietnam.)

The whole reservoir area of roughly 65 percent (1.4 million hectares) in China is cultivated and maintained as fish ponds out of a total of two million hectares. The fingerlings are stocked and mature fish are taken after a period of growth. The species employed are the common carp, the grass carp, the silver carp, the bighead carp, the mucus carp and the Wuchang bream in different densities and species ratios (Beveridge & Phillips, 1988). In China, reservoir fisheries rely almost exclusively on the regular release of vast quantities of fingerlings as the Chinese carp does not spawn out of its own habitat. Thus supplied fish account for more than 90% of catches from reservoirs in many situations (Barman et al., 2013).

A wide selection of species including tilapia, mullets and carp, normally cultivated on tightly maintained polyculture systems, are currently stored in Israel in numerous storage reservoirs (10-40 ha, 5-7 m deep), initially established as irrigation water. The country's major freshwater resources are massive perennial reservoirs and minor seasonal tanks. The carp (Indian, Chinese and common) are regularly stored in fingers manufactured in government-run hatcheries around the country. As Indian and Chinese carps are not spawning in reservoirs/tanks, they must be stored regularly to support the fishing industry. There is consequently considerable pressure on the State to speed up its seed production scheme to meet large fingerlings requirements. It is noteworthy to note that the Sri Lankan Ministry of Fisheries is mainly responsible for producing carp fingerlings for distribution between various reservoirs and seasonal tank, unlike in the Philippines and Thailand, where fish seed production is largely a private sector company.

In Cuba, the majority of freshwater fish are produced by the seed of the small to medium-sized reservoirs in this country (FAO, 1986).

Little reservoirs of Tilapia and *Clarias* spp. are largely supplied in Africa. *Tilapia melanopleura*, *T. macrochir*, *T. andersonii* and *Haplochromis* spp. are the most important species in Zambia for the storage of reservoirs. *Oreochromis mossambicus*, *O. macrochir*, *T. rendalli* and many exotics including black bass and rainbow trout are the most often used foodstuffs for dam stocks in Zimbabwe.

## 2.11 FLOODPLAIN FISHERIES MANAGEMENT

With a vast network of swamps, lagoons, lakes and ponds, the tropical river plains provide enormous possibilities for a rise in fish yield, through improved and more effective management, whether employing catching or culture techniques or combining both.

Floodplain fishing is a seasonal sport based on the length and volume of flood filling the lowlands. Consequently, the management of aquaculture and/or culture-driven aquaculture floodplains needs to be timed to correspond to the seasons and the water level in the floodplain.

The easiest way to manage/intensify the floodplain is to connect the main river channel with the neighboring floodplains by cutting a canal through the levy. Water and fish are swallowed up through the canal when the water levels rise and catcher by various gears when the flood waters are returning to the main river (Benetti et al., 2021). In Africa, where there are huge riverine systems and related floodplains, floodplain management is frequently practiced. It is thought that practice began in Sudan in the 1940s.

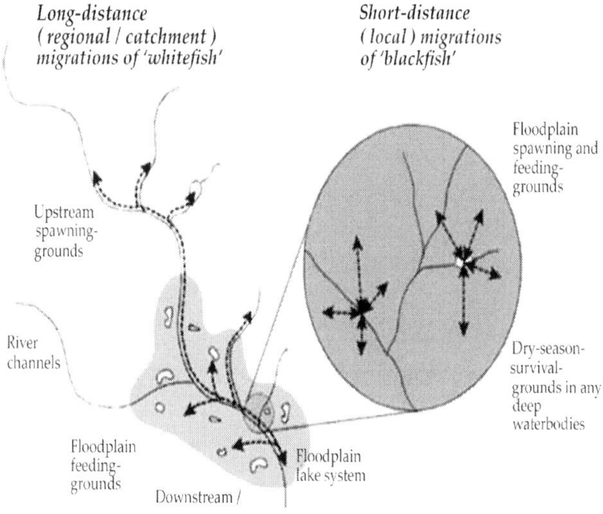


**Figure 2.11**: In the lagoon, the engineering work also serves a range of goals such as providing greater fishing surfaces and the passage by the neighboring

lagoons of fisheries and shellfish. Interconnecting waterways also provide better access to nursery regions for larval fish and shrimps. Sedimentation reduction and circulation enhancement are other aims of internal canals Alternative migration patterns of riverine 'whitefish' and 'blackfish'.

*Source: http://www.fao.org/3/x1357e/X1357E03.htm*

Natural levees were erected on the Ubangi River in Zaire, with open canals that allowed fish to access surrounding swimming pools and subsequently closed down with winding to make a huge fish trap. The pools can be stored so thickly, that the fish must be fed.

In Benin, rectangular fish ponds frequently perpendicular to the river have been excavated to improve the habitat of the fish and easier their catching as the flood waters falls. The annual flood fills the ponds and walls the entrances from the main river, fishing the swimming pools in order to clear the vegetation in the ponds and the barriers advance inland to concentrate the fish before they are captured toward the conclusion of the dry season (Bostock et al., 2010).

## 2.12 BRUSH PARKS OR ACADJAS

Another type of floodplain management involves the establishment of brush parks or acadjas, as has been done in various West African countries, as well as in Cambodia and Myanmar in Asia.

In brush parks vegetation is placed in shallow waters of floodplains to attract fish. Before capturing the resident population, the 'park' is later netted. In more simple circumstances, swimming vegetation can be put in the water in search of food and refuge to attract fish and fished in a couple of days. The floating vegetation is in more advanced circumstances encircled by poles or branches driven into the floors of the river or lake. Construction happens following the annual flood, with projects fished at intervals ranging from several weeks to several months according to conditions. Yields of up to 28 tons per hectare per year have been reported for brush parks in Benin's Ouémé River, without influencing other fishing methods in the vicinity.

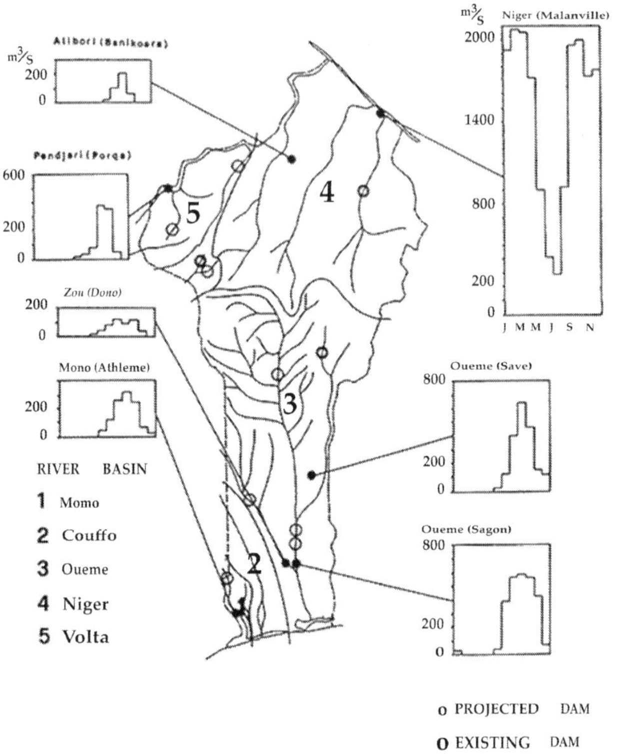


**Figure 2.12**: The country can be divided hydrologically into five principal drainage areas.

- Northward Benin borders Niger across around 120 km and on the central plateau of Benin, three of the river's major mid-range tributaries (the RivièresAlibori, Sota and Mekrou) rise torrentially towards the Niger.
- The river Pendjari, which is a tributary of the VOLTA, drains the western slopes of the Atakora Mountains and passes through Benin for almost 400 km prior to entering Togo, covers an area of around 275 square kilometers at the highest floods.

The south-flowing rivers:

- The Mono (360 km),
- The Couffo (240 km), and
- The Ouèmè (700 km), arise on the central plateau and, below torrential upper courses, form broad fertile floodplains in the coastal zone. The floodplains terminate in a series of extensive

lagoons. The greatest of the floodplains, that of the Ouèmè, extends over 2 000 km$^2$ at peak floods.

*Source: http://www.fao.org/3/T0360E/T0360E02.htm*

The management of the flood plain is thought to be an interesting alternative for increasing production of fish by large still largely under-used water bodies because of the low costs associated with the establishment of brush parks and other facilities for modifying the flood plain, especially in areas of the world where aquaculture per se may not be reasonable because of the current practice.

In conclusion, in the past 20 decades, the aquaculture business has expanded in several countries worldwide from an experimental/pilot phase to a fully established, vital subsection of the fishery, but especially in Asia, from where the majority of international production is produced (Corsin et al., 2008).

However, given this aquaculture, it is not and should not come as a surprise:

- It can supply the masses, including those countries without marine access, with an affordable supply of animal protein in rural areas;
- may be created on land which, in combination with other farming systems, is no longer viable for agriculture or on land;
- It can be used on a small scale, at low costs and using family/community labor, or on a big scale at high costs and with the use of more and fewer machines, and satisfy the goals set for both of these;
- may supplement fishing by enhancing fish output in open waters such as coastal lagoons and man-made lakes;
- It may offer a viable socio-economic alternative to catch fishing particularly in municipal overfished waters;
- It may improve fishing resources in large water bodies with natural productivity indicating declines in overexploitation or damage to the environment.

There are a variety of possibilities available in aquaculture. It can be enormous or little. It can be done in fresh, brackish or salty conditions on a terrestrial site. It can use wide, semi-intensive or intense approaches for a wide range of cultivars with various economic values.

Its goods can be sold either domestically or internationally fresh or processed.

This flexibility is however conceivable, only if:

- The product market is guaranteed;
- The species picked can be cultivated and meets the usual criteria for the good species candidate;
- The location selected is suitable for species cultivation
- Technology and type/method of cultivation available for the selected species;
- adequately designed and constructed production and support facilities;
- The operation is efficiently handled in order to ensure proper use and availability of the various production inputs at the proper time;
- Able to administer and operate the company are properly trained and skilled workers;
- There is appropriate finance and/or development and/or operational credit available.- the location selected is suitable for species cultivation (Bondad-Reantaso et al., 2005).
- Technology and type/method of cultivation available for the selected species;
- Adequately designed and constructed production and support facilities;
- The operation is efficiently handled in order to ensure proper use and availability of the various production inputs at the proper time;
- Able to administer and operate the company are properly trained and skilled workers;
- There is appropriate finance and/or development and/or operational credit available.

Evidently, only through proper management based on knowledge of the traditional environment and biological processes associated with the cultivation operation and on the existence/availability of certain vital infrastructures such as fry, feed, labor, funds and a market can the success of aquaculture be achieved.

**Figure 2.13**: Aquaculture fishing.

*Source: https://www.conserve-energy-future.com/aquaculture-types-benefits-importance.php*

At the domestic level, the government must ensure that fish farmers are given the required support in their production efforts, especially through the following measures:

- Appropriate infrastructure and services for the market (e.g. farm roads, fishing ports, processing plants, ice-cream plants, cold storage);
- Excellent services of training and expansion;
- On-going efforts in research and development (Bondad-Reantaso et al., 2005);
- The provision of financial incentives, for example, tax credit, free import of required machinery / equipment and the provision of supplies and materials, and credit / financing support.

Improvements in cultivation systems, extremely productive strains, highly enhanced feed formulations in the intensive farming systems, highly improved techniques in fry hatchery and expansion of production areas have all made a significant contribution in many countries of the world to the exceptionally quick development of aquaculture. In the years to come, this quick expansion of the sector will certainly continue if not exceeded when the enhanced processes will be refined, the feed will be even improved and larger yields will be produced genetically. And as highly developed aquaculture nations support others who are less developed than them, the global vision of fish on all tables can still be achieved.

Chapter 3

# CAPTURE OF JUVENILES IN THE WILD AND GROWING OUT IN CAPTIVITY

## CONTENTS


3.1 Introduction ........ 64

3.2 Economic and Social Impacts of Capturing Juveniles from the Wild ... 67

3.3 Selection of Species ........ 74

3.4 Growing Out of Juveniles in Captivity ........ 78

## 3.1 INTRODUCTION

Total aquaculture production has expanded tremendously and contributes to the world's supply of fish for human consumption in an expanding significant way. In recent decades, this growing trend is expected to continue. The sector is expected to contribute in a far better way to food safety, poverty reduction and economic development by generating, by 2030, 83 million tons of aquatic products – with minimal environmental impact and maximum social benefits; an increase of 37.5 million tonnes. Aquaculture is a multifaceted sector with different fish production systems (Waagbo & Remo, 2020). One of these tactics is the collection of wild individuals, whether as a flourishing plant whose eggs grow in swamps or cages or as early phases of history for constrained and regulated growth. The Food and Agriculture Organization of the United Nations (FAO) has designated this farming approach as catch-based aquaculture (CBA) and is employed worldwide in a range of marine and freshwater species, which are of significant environmental, social and economic significance.

In comparison to aquaculture, captures of aquaculture which regulates the complete reproductive cycle of farmed animals have various benefits and disadvantages. The system does not depend on the reproduction and breeding of farmed species being controlled. Therefore, hazards and breeding programs can be produced for species of high value or for species that have an easily available natural environment. However, the lack of domestication opportunities for wildlife is the main disadvantage, because even long-term improvement in genetics is not conceivable. This method of aquaculture is used for marine fin-fish species of great importance, for example, tuna, which demand high protein diets and robust cultivation plants. It is also utilized, however, with low-value fish species, sometimes cultivated with minimal inputs in small ponds or cheap farms. The first offers economic opportunities, but requires considerable investment and infrastructure, while the latter provides food security and a supplementary source of income for rural populations. In light of economic viability, the intelligent use of natural resources and the environmental impact in general, all kinds of CBA must be examined.

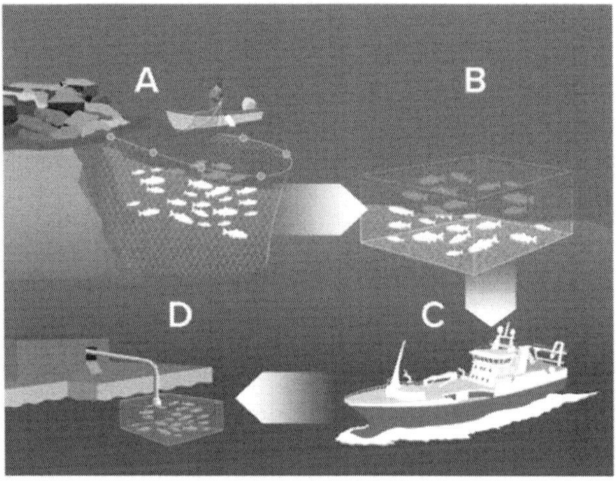


**Figure 3.1**: Aquaculture is a multi-billion-dollar enterprise that employs a variety of fish production methods. One of these strategies is to harvest the wild people in confined and controlled conditions, from extremely early stages in their life cycle to large, mature adults.

*Source:* https://www.sciencedirect.com/science/article/pii/S1466856420303805

While it is impossible to measure the scope and breadth of CBA practices, an estimated 20% of the productivity of marine aquaculture is estimated at 1.7 billion USD per year. The cultivation of many species of freshwater also relates in whole or partial terms to fries obtained in the wild because the supply from hatcheries does not satisfy demand or because farmers think that the quality of hatchery-produced seed is lower than that of wild seed. CBA's primary concern is if the seed fishing impacts both the wild stocks of the target and non-target species. Although there are usually little data on this subject, some governments have tried to ban or restrict such fishing in some way (Tucker & Hargreaves, 2009).

The collection of natural resources for CBA has environmental concerns that need to be addressed. Most controls on fishery management are limited in size for harvested species and spawning adult harvesting is often restricted. Targeted farmers are early stages of life history and adults who are ready to spawn, but are not covered effectively by existing legislation. To determine the sustainable use of CBA must address the impacts of these aquacultures on wild populations "targeted" and on their related non-target species and the surrounding ecosystem. Many scientists, government specialists, aquaculture suppliers and manufacturers, aquaculture product retailers and environmental and social advocacy or key stakeholders have noticed these sustainability challenges in aquaculture development. Numerous international and national and intergovernmental meetings have decided that the difficulties which impede sustainable aquaculture development at the global, regional, national, and local levels must be dealt with and resolved.

Aquaculture and the necessity for international cooperation to promote its sustainable growth, as well as its potential contribution to the improvement in many rural areas, have been discussed by FAO and the Committee on Fisheries (COFI) on numerous occasions. The FAO Ministerial Meeting in 1999, as well as the inaugural and second Meetings of the COFI Sub-Committee on Aquaculture, all emphasized the importance of the international aquaculture industry stepping up its efforts to promote more responsible and sustainable aquaculture production methods.

**Figure 3.2**: Special demands on animal welfare are made by capture-based aquaculture. Therefore strategies and procedures created in regard to their gentleness ought to be documented.

*Source: https://nofima.no/en/nyhet/2012/10/national-centre-for-capture-based-aquaculture/*

A Trust Fund agreement has made provisions to tackle selected major sustainability questions in global aquaculture operations and progress through the 'Towards Sustainable Aquaculture – Choosing Issues and Guidelines' initiative implemented by the FAO and supported by the

Government of Japan. The theme area of 'use of wild fish and fishing resources for aquaculture has been recognized as a priority of target action in accordance with the proposal of the FAO Committee on Fisheries and Aquaculture Sub-Committee at its first two sessions.

The program also concentrated on the collection and synthesis of information available on the above-mentioned and other specific objectives and, depending on the evidence analyzed, offered possible management systems and options for targeted responses to the specific concern, including limitations and issues. The results provided by the project will help Member States of FAO to promote and apply the provisions of the Responsible Fisheries Code of Conduct (CCRF). The particular purpose of this project sub-component is to make a contribution to the effective and comprehensive cultivation and preservation of natural aquatic populations at the international market by successfully implementing ecosystem approaches to fisheries with minimum disruption to responsible and sustainable fishing resources (Tacon & Halwart, 2007). In order to attain such a purpose, the FAO hosted a workshop on responsible use of wild fish and resources for aquaculture in Hanoi, Vietnam, in October 2007 in which it started the creation of a number of technical guidelines. These once accessible technical guidelines are intended to assist policymakers in the development of laws and measures which take account of the use and preservation of aquatic resources.

## 3.2 ECONOMIC AND SOCIAL IMPACTS OF CAPTURING JUVENILES FROM THE WILD

Aquaculture is viewed as a viable option for meeting present and future needs for marine products. Many aquaculture processes, however, still require significant modification in order to make them more sustainable, minimize their reliance on wild fishing sources, and prevent hurting aquatic environments. The continuous dependency on small and small value or by-catch, usually referred to as garbage, for feed and environmental consequences as for the flow of nutrients into coastal waters comprise major obstacles for the creation of effective aquaculture. The capturing of early life-history animals (i.e. larvae, fry, fingerlings and young) or "seeds" from the wild for growing to market size in aquaculture facilities is another restraint. A substantial restriction on aquaculture expansion could be sustainable access for fry and fingerlings. The "catch-based aquaculture" approach has been described as an overlap among fish capture fishing and aquaculture. In FAO statistics, this operation is reported instead of capturing the fisheries, while it is not a hatchery, but is dependent on wild seed supplies.

Due to the market requirements for certain high-value species, the capture-based aquaculture has grown which now cannot conclude their life cycle at a commercial level. Furthermore, there is still much poorer and less dependable survival of larvae in hatchers in the hatchery production of a number of cultivated species. Fingerlings and fry supplies of many wild cultivated species have decreased, likely to be caused by overfishing, the degradation of habitats, harmful fishing practice, pollution, high demand for exports and high mortality of captured fry animals. Such capture-based aquaculture includes tuna (*Thunnus* spp.) in the Indonesians and Philippines; eels (*Anguilla* spp.) in North America, Australia, Europe, and Asia; grouper (*Epinephelus* spp.) in Asia; and reef fish settlements for the trade in marine water-powered products in Spain, Croatia, Canada, the Japan, and Australia; the Philippines, Turkey and Morocco.

This research examines the social and economic implications of fish farming and, in particular, the capture of ELP creatures from nature as seeds in marine and freshwater cultivation. The issues and implications emphasize the overlap between capture fisheries and aquaculture in fish farming. Species selections for cultivation reflect demand in local and worldwide markets and consumption tastes and preferences, and are the driving force behind the growth of a capture-based aquaculture business (Sivasankar et al., 2017). Other economic problems include expense and profit (privileged and social) and market channels. Social questions include jobs, livelihoods, infrastructure development, and rights to property, crises, technologies, culture/traditions, ethical views and public engagement. This paper will address these difficulties and their impact. The aquaculture production system known as capture-based aquaculture is characterized and technically different dependent on the place and type of the world. As demand and technologies develop, the utilization of this aquaculture approach is continually evolving.

**Figure 3.3**: For intervals of several hours to several months, the fish are preserved alive. Our investigations reveal that the quality is substantially greater by using short live storage on-board trawlers than by slaughtering fish straight after being pulled on board.

*Source:* https://nofima.no/en/research/capture-based-aquaculture/

## 3.2.1 Using of wild seed for the capturing of Juveniles

The wild-caught seed aquaculture of a number of marine and freshwater species is performed worldwide. Because of lack of reporting and statistics, the size or percentage of aquaculture output in the freshwaters and maritime environments that are dependent on the capture of ELP animals from wild animals is extremely difficult to determine accurately. FAO has projected that, from capture-based aquaculture with a value of $1.7 billion, 20% of marine aquaculture is produced in 2006. Many species of freshwater additionally have partial or complete cultivation of wild seeds as the supply from hatcheries is not adequate to satisfy demand or as their quality is regarded to be below that of wild obtained seed.

There has been no estimate for aquaculture productivity based on freshwater capture. Reports show that the usage of hatchery seeds for many species is growing as technology improves and the supply of wild seed is decreasing. The fact that many aquaculture species are not supplied with reliable, natural seed was significantly hindered by continued growth and development. The shifting nature of aquaculture seed supply, from wild to hatchery-produced, contributes to the challenge of establishing a precise assessment of aquaculture reliance on the wild seed.

A few studies and publications from around the world provide estimates of individual species, illustrating the on-going reliance on seeds obtained in wilderness, even when exact data for the scope of catch-based aquaculture are not accessible. It was stated that all the seed is taken from the wild in Cambodia in some freshwater species such as omnivore catfish (*Pangasaianodo hypophthalmus*) and carnivorous giant snakeheads (*Channa micropeltes*). One scholar predicted in the Philippines that while hatcheries have become a growing source of milkfish, only about 15% - 20% of the demand was supplied by hatcheries (Subasinghe, 2017).

Even though many bivalve species are regularly generated in hatcheries, wild spat generation often dwarfs the output of hatcheries. It is estimated that up to 95% (Mytilidae) spit has been harvested from the wild; about 15% to 20% (Osteridae) and about 50% (Scallops) are obtained from edible oysters (Pectenidae). Spiny lobsters are produced largely from the catching of wild juveniles (*Panulirus ornatus*), in Asia, (Vietnam, Thailand, Singapore, Philippines, Myanmar, Malaysia, India, and China).

Despite the availability of hatchery-reared groupers, wild-caught juveniles continue to be the main supply of seed for these species' aquaculture in Asia. Adult grouper collected in the wild account for about half to two-thirds of the area supply. Malaysia, Thailand, Indonesia, and the Philippines, are major sources of wild-caught grouper. Wild-caught grouper is also available from China, Papua New Guinea, Myanmar, Vietnam, and Australia. Micronesia (Federal State of), the Maldives, the Solomon Islands, Kiribati and Fiji are among the new supply sources in the Indo-Pacific.

*Macrobachium rosenbergii* or huge freshwater prawn in Vietnam is becoming a growing farm species, which is an indigenous species in the Mekong delta. This kind of cultivation is mostly focused on wild seed from rivers and other freshwater sources, notably in rice fields. In Vietnam, the main source of sand goby seed is the wild. Although not for food, ornamental fish production in a number of nations is an important component of global aquaculture. Freshwater species are mostly concerned with the aquaculture of ornamental fish. About 90% of the ornamental freshwater fish are caught.

Although marine ornamental products capture better prices, they have considerably less advanced breeding and culture. In the pet sector, about 100 out of 800 marine species are grown on a routine basis and approximately 21 of them are commercially practicable. Many key farms still rely, as can be shown, on the usage of wild creatures as seeds. The rise of the capture-based aquaculture business has been fuelled by market forces. As is the case with

Bluefin tuna capture-based aquaculture, market requirements are essentially defined by customer tastes and conventions. The great demand for this high-quality species was fuelled by the production of traditional Japanese "sashimi," leading to the development of capture-based aquaculture.

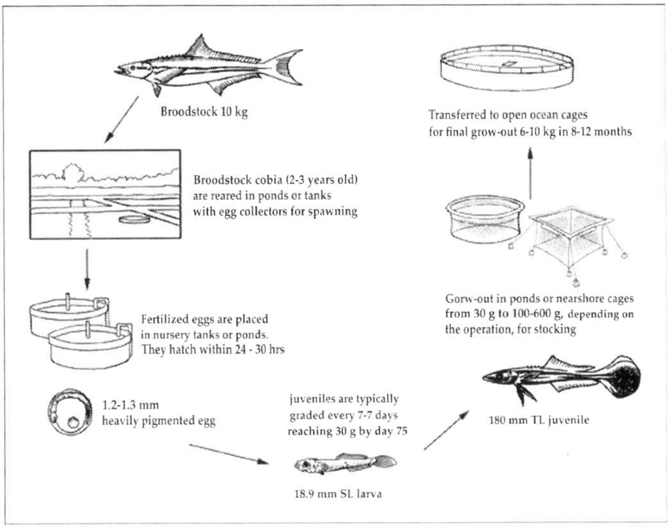


**Figure 3.4**: The aquaculture production of cobia in pools may lead to water quality problems and excess nutrient loading in the effluent, depending on the pond, crop density, feed rates, and water exchange rate. Located on the shore, large-scale cobia production at different phases of life could affect any coastal environment and necessitate monitoring of the production outflow.

*Source: http://www.fao.org/fishery/culturedspecies/Rachycentron_canadum/en*

To be lucrative and effective, all businesses should keep good documentation, and capture-based agriculture is no exception. To understand the achievements, failures, and profitability of any venture, meticulous record-keeping is required. The growth of this activity has been ensured by strong profitability combined with significant market demand. However, the rise in capture-based aquaculture has a variety of major and diverse socioeconomic consequences, not all of which are positive.

In local, national or worldwide markets, the selection of species to be grown depends on their admissibility. Based local, ethnic, economic, and cultural traditions, capture-based aquaculture systems can change entirely. The organization of fisheries, farming operations and utilized technology are partly responsible for cultural and ethnic heterogeneity, as well as for economic inequalities (Stentiford et al., 2020). The absence of adequate technologies and limited investment in capture-based aquaculture activities in Asia may be restricted. For instance, groupers are often small and handcraft-based in capture-based aquaculture, but tuna is high-tech and demands significant capital investment. In conjunction with local operators, international enterprises generally invest in the latter form of culture.

Depending on the country and species, socio-economic concerns differ. Capital requirements for grouper farms are low, community and family-based, yet income can influence the rural/local economy significantly. In the local economy, from "seed caught" to feed traders and suppliers, various earnings might be made. Tuna farming has substantially smaller effects on the local economy when handled by foreign businesses since the seeds of fish are gathered up in higher seas and the goods are sold on worldwide markets. Tunas tend to locate the revenues as wages, while international corporations generate actual profits.

It is not always the social benefit to exploit a common resource. The danger of the density of the advantages into a few hands is often exacerbated by the lack of an adequate regulatory framework. It can cause conflict between fishers, farmers and other resource-users (e.g. tourism and shipping). Consequences are not always foreseeable, whether they are favorable or negative, because capture-based aquaculture is a new industry with a little history and its features are species-specific. As a result, gaining a broad perspective across the socioeconomic spectrum is quite challenging. In the short run, conflicts with other competing resource users and sectors are unexpected, necessitating expert sector analysis (socio-economic, environmental, marketing, etc.).

Fish farming now represents 8 times the economic worth of all regional fishing in Spain's Murcia region. This is due to a quick response to the demands of the Japanese market and the establishment of strong ties with Japanese enterprises. In 2001 there was a development of new markets in the USA and in Europe; exports to Europe were stated to have risen by 50%, 76% to America and 3% to Asia. Croatia Bluefin Tunas is operated by 6 commercial enterprises that use 9 rental locations. Approximately € 50 million is the economic advantage to the fishery sector from captured aquaculture.

**Figure 3.5**: While it is unlikely that post-larvae capture and cultivation will meet the demand of the ornamental trade for all tropical fish, the capture and cultivation of post-larvae can create large niche markets for eco-labeled specimens and provide coastal villagers sustainable economic advantages from coral reef resources.

*Source:* https://fish.cgiar.org/news-and-updates/news/youth-fisheries-aquaculture-and-value-chains-study-and-foundation-building

Social benefits frequently are strongly linked with economic advantages. New industry development is capable of creating employment. In the case of capture-based aquaculture, these jobs generally originate in remote coastal towns with scarce jobs. The fisheries business also has a relationship. Another fisheries-related industry is a welcome choice for the present qualified workforce with a steady reduction in fishing possibilities. Youth drift to cities can also be reduced if jobs are offered in a newly vibrant sector that is capable of generating high pay (Rimmer et al., 2013).

The Mediterranean tuna fishing sector has been modified by bluefin tuna. Fishermen's tuna prices soared and operating processes were modified (including fishing seasons and areas, as well as fishing operations). The need for bags, used for the collecting of live bluefin tuna, is also on the increase. The development of bluefin tuna aquaculture in Croatia has had a considerable influence on the fishing sector. After Southern bluefin tuna fattening techniques had been successfully adopted by Australian immigration, the practice spread quite quickly.

Within just a year his fishing boats grew to 30 from 19 (1999) to ensure supply to tuna farms. In France, the same phenomenon occurred, mainly as a result of the increasing volume of tuna catch farms in Spain, with the number of specialized blue-fin tuna fleets increasing from 21 to 32 boats between 1995 and 2000. Increasing numbers of baggage carriers (including in other countries in the Mediterranean) contribute to fleet modernization, enhanced footwear efficiency, speed, increased safety, capture system and comfort for crews.

Many fishermen are becoming active participants in tuna farming, either as suppliers or as tuna growers. Some fishing vessels are used to catch small pelagic fish that are utilized as caged tuna feed. Trawlers have been extensively incorporated into tuna farming operations in Croatia, either to carry live fish or to distribute feed to the farms. Tuna farms provided an essential source of employment in the highly depopulated Croatian islands, with 300 farm jobs generated. The bluefin tuna industry in Spain's Murcia region employs 500 people directly. The employees are youthful, ranging in age from 25 to 35 years old, and the bulk of them work as scuba divers at sea.

In Australia, catches for bluefin tuna (purse sein system) are less than long-lining in the High Seas: A$3.50/kg vs. A$22/kg. Catching costs are below the cost for catching and wasting resources (B. Jeffriess, pers. comm. 2002). Australian catch-based tuna aquaculture, due to its work intensity (350 direct and 700 indirect jobs), was significantly more expansive in terms of operations, infrastructure requirements and exports of fresh frozen and chilled foods. Bluefin tuna farming has contributed to the social cohesion of the tuna sector as harvesting crews of wild tuna remain gone for quite lengthy periods. Better working conditions than aboard fishing vessels (weekends on shore, regular salaries, regular hours). In Australia, for instance, where a special unit of divers cross tuna through a sweep net into a small area of the cage and then catch Bluefin tuna immediately by hand, the tuna weight of tuna is average at 50 kg (Rahman, 2014). This skill growth is crucial for harvesting operations. This approach is very significant and it is crucial that divers who are not too stressed be experienced because this has an impact on the quality of the product.

## 3.3 SELECTION OF SPECIES

Its potential brand value, growth rates and the ability of a species to operate under circumstances of culture are the main criteria that define the acceptability of a species. Even more crucial for an aqua culturalist than biological elements is economic concerns for the selection. Carnivorous types usually require a high diet of protein and are consequently regarded as more expensive to rear, however, expenses are mostly dependent on the local availability and prices of the necessary feed. However, the more carnivorous species have higher market prices in compensation for high feeding expenditures.

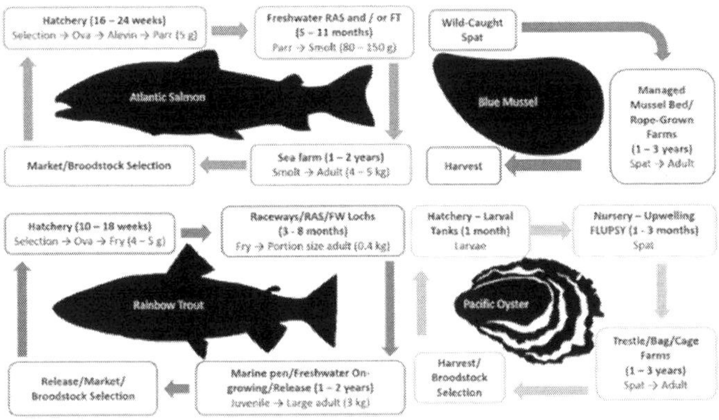


**Figure 3.6**: Early adoption and extensive adoption of well-managed selective breeding operations, which compounded increases in growth performance and qualities such as disease resistance, were vital to this achievement for Atlantic salmon. However, genetic technology differs enormously between species and these technologies progress quickly and become more accessible and affordable.

*Source: https://onlinelibrary.wiley.com/doi/full/10.1111/raq.12553*

In general, such species have increased market potential, thereby attracting significant investments. For most aquaculture operations, this is particularly important for catch-based aquaculture, as most of the species cultivated are carnivorous. The species groups of mollusks (scallops, oysters, mussel), crustaceans (shrimp, crabs) and fin-fish are utilized in catch-based aquaculture (eels, grey mullets, milkfish, yellowtails, groupers, rabbitfish, tunas). Four target categories of species are included in this report: yellowtails, tunas, groupers, and eels. The latter have been chosen for their importance in captured aquaculture because they are rapidly growing and have significant market demands; their culture and technology advancements are also increasingly interested to support their upbringing.

Groupers are famous South-East Asian fish and, due to their quick growth, effective feed consumption, high market values, and lower wild available resources, can become significant aquaculture species. In the last two decades, demand for groupers in part of South-East Asia has surged sharply. Live groupers vary according to species. Price varies between USD 8-31/kg in Hong Kong, for instance (INFOFISH 2003). Groupers are also good aquaculture prospects for gastronomic reasons; in many parts of the world, they are considered to be one of the best quality fish products (Roubach et al., 2019).

Another great species for diversifying agricultural fish products is amberjack and yellowtail due to their high growth rate and strong captive performance (low mortality, acceptance of food). In particular, in Japan, the Yellowtails market is good; it has developed due to capture-based aquaculture output over the previous 30 years. One of their most noteworthy features is that they may be processed and marketed as a variety of products, such as complete items, steaks, fillets, and so on. This is one of the few examples in which farm fish are widely regarded as of superior quality to sea-caught fish and have a far higher price on the market.

The culture of other species from the same group, especially the bigger amberjack (*Seriola dumerili*) and the gold striped amberjack (*Seriola quinqueradiata*), is enhancing the traditional production of the Japanese amberjack, commonly known simply as "yellowtail" (*Seriola quinqueradiata*) in Japan (*Seriola lalandi*). *S. lalandi* is highly prized in Japan for fresh consumption as "sushi" and "sashimi"; *S. dumerili* has cultural potential in the Mediterranean (France, Greece, Italy, and Spain). To meet rising export demand, South Australia has begun harvesting *S. lalandi*. Market prices for Japanese amberjack, gold-stripped amberjack and larger amberjack, in Japan, mostly for "sashimi" consumption, range from 1 200 to 3 000 yen per kilogram (Nakada 2000).

Eels are a popular commercial aquaculture species, and they are regarded as a delicacy in many places. Western Europe and Japan have traditionally been the areas with the biggest demand. Eels can cost up to US$ 32 per kilogram, more than twice as much as good-quality shrimp. In Japanese restaurants, where eels are considered a nutritious diet, a serving of eel might cost anywhere from $20 to $32. Eel aquaculture has evolved into a highly specialized sector, with the global demand for eels exceeding 200 000 tonnes per year.

The bluefin tuna is the fourth species we've chosen, and it's becoming increasingly popular among farmers. The practice of capture-based tuna farming has exploded in popularity during the last few years. *Thunnus maccoyii* in Australia, *Thunnus thynnusorientalis* in the North Pacific, and *Thunnus thynnusthynnus* in the North Atlantic and Mediterranean, are the three main populations studied. The market demand for "sushi" and "sashimi" items in Japan has fuelled these advances, and many of the initiatives are collaborative ventures between local fishermen/fish farmers and Japanese firms.

**Figure 3.7**: CBA also affects the stressors introduced to fish during capture and the duration of these effects increases considerably from hours and minutes for traditional fishing to days and months for CBA and also in terms of adaptability to new surroundings.

*Source: https://www.ifpri.org/sites/default/files/gfpr/2015/feature_3087.html*

In the Mediterranean, Australia, Mexico and Japan, there are manufacturing centers. This active aquaculture sector is now established and is the most profitable, despite its hefty development expenses. The prices on the Japanese market for Bluefin tuna justify the premium given. Prices for *Thunnus obesus* (big-eye) are usually five times that paid and ten times that paid for *Thunnus albacares* (albacore) (Doumenge, 1999). A cultivated Bluefin tuna intended for the raw fish market ('sushi' and 'sashimi') has become an excellent product, and is particularly ideal for 'sushi' with its increased fatty content. Catch-based aquaculture fish not only produce considerable weight gains but also enhance the content of fat, which makes them more marketable. 1 kg high-quality tuna can be sold on the Japanese market for up to 600 US dollars.

Capturing-based aquaculture involves the comprehensive realization of each species' behavior, habitat and general environmental requirements and their reproductive biodiversity, nutritional requirements, larval and young physiology, cultivation systems and the availability of 'sowing' products. Furthermore, the susceptibility of each species to disease under traditional circumstances requires research. However, all will be wasted unless the market prices sustain additional investment and future research, despite technological advances in capture-based aquaculture.

## 3.4 GROWING OUT OF JUVENILES IN CAPTIVITY

In those instances, where it has been decided that existing or projected CBA activity is appropriate, guidelines are provided in the following section to ensure that the aquaculture operations are carried out after CBA has collected wild material. This part covers the general questions relating to the handling and transportation of live materials, culture and cultivation, documentation and usage of fish fodder. While these problems related to post-fishing also relate to HBA and have been treated similarly in other recommendations, they are of particular or unique importance to CBA.

Traditionally relevant issues that are not immediately and especially relevant to the CBA wild capture component are not treated by this document, however. A number of scientific guidelines for FAO, for example on the health planning of culpable mobility of live aquatic life (FAO, 2007) and on the management of genetically modified resources include an excellent coverage of crucial problems varying from best farming techniques, waste from agriculture, genetic factors, health considerations, site choice and water preservation (FAO, 2008b).

After the catch of live materials, the survival, health and wellbeing of seeds and brood-stock are vital as they transit from capture to cultivation. Initial capturing and grading and sorting processes can lead to large levels of mortality during transportation from the capture region to the aquaculture plant and as part of a process of regeneration or acclimation upon arrival at aquaculture plants. The apparent effects of major mortality at the present level on the economic feasibility of CBA operations on the health and welfare of animals warrant scrutiny. Elevated levels of mortality from harvest, post-harvest or pre-aquaculture motivate additional fishing efforts to restore the waste products, leading to greater environmental and ecological efficiencies. Mortality data are necessary if live material from catch to aquaculture is to be managed and improved.

**Figure 3.8**: In all stages of the Mahi aquaculture, researchers have made steady progress and agree that difficulties can be addressed if adequate resources are used. The technology is in place, but the successful renewal of efforts to build commercial aquaculture for the species is eventually determined by market forces. If there is good demand for the market and favorable prices, mahi may be increased.

*Source:* https://www.aquaculturealliance.org/advocate/mahimahi-aquaculture-revisited/

Even if target animals survive at a high rate, sub-lethal strain, trauma, competitive interactions among animals, and other variables might damage their health and welfare. Inadequate health and wellbeing may limit the live material's ability to survive in confined farming facilities or jeopardize its ability to achieve optimal development or quality while in captivity, posing a financial risk to the CBA operation (Rodrigues et al., 2012).

Best approaches to decrease death and maximize animal health and welfare will vary between various species and among the phases of life for live animals. A range of problems, technology, methods, equipment and approaches may be included in best practice. Suitable handling, transportation and procedures, for example, will have to take into account elements such as tank size, sampling, conditioning, transfer, recovery time and inspection. To optimize the health and well-being of the living material, adequate monitoring and water exchange will be required to preserve water quality.

In accordance with projected biomass and volume, the stocking density must be within the adequacy of the species and the size and condition of animals. To improve CBA as a whole, it is vital to transfer the best practices and lessons acquired from mortality and optimization of health and welfare. Management of live CBA material, including seed and brood-stock, should aim at minimizing mortality, addressing animal welfare concerns and making optimum use of natural resources and undermining waste. In addition:

- Best procedures to limit mortality and maximize the health and well-being of the live material from capture to confinement should be established and executed in all phases.
- Methods to facilitate surveillance and reporting should be documented.
- In order to detect and address problems that affect survival, safety and welfare of live materials, data on fishing, post-collection and pre-aquaculture mortality should be collected, assembled and analyzed.

**Figure 3.9**: Apart from selling totoaba in restaurants and retail outlets throughout Mexico, they have supported the restoration of wilderness from the beginning, releasing to their natural environment thousands of Totoaba juveniles and working with local schools to educate the coming generations on the significance of the conservation of wildlife. So far over 100,000 young people have been released into the wild by Earth Ocean Farms to safeguard this unique Mexican species.

*Source:* https://thefishsite.com/articles/can-aquaculture-save-a-species-from-extinction-totoaba-mexico

Capture stress and the transfer of living materials, including seeds and brood-stock in the field, impair the health and welfare of CBA. Increased density and water quality, and inadequate management, together with stress, may increase the incidence and infection of disease and may become an issue in animal welfare, and transport and storage in restricted spaces or holding facilities. In order to maintain optimum health, aquatic organisms may be required before they arrive at the farm facilities to minimize stress and prevent or eliminate sickness or infections in the collected stock. Credible information on the most suitable and successful treatment items and methods for individual stress, sickness, and infection or species demands is not often available. Many regulations on consumer protection and biosafety problems have been introduced in recent years (Rosa et al., 2012).

Often, substances such as anaesthesia (e.g., phenoxyethanol), antifoam, antioxidants, and antibiotics agents are restricted, as are all goods which are effective in decreasing fish stress and in treating and preventing diseases or infections which are a concern for human health. The bioaccumulation of these drugs in living materials, safe levels of human intake and potential transmission to consumers via aquatic food products are not well known. Procedures should therefore be cautious for their utilization. Authorized veterinarians should apply pharmaceutical therapy. During transport and handling from capture to captivity, they should deal with stress, diseases or illnesses, be appropriate and not adversely affect human safety. Specifically:

- The most adequate and effective treatment for specific stresses, diseases, infections or species situations should be obtained.
- The best information should be available.
- The drug treatment regimens should be based only on approved and prescribed doses of the drug therapy agents.
- Data should be collected, assembled and evaluated on the occurrence of disease and infection and on the pharmaceutical therapies.
- Treatments by adequately qualified personnel should be diagnosed and supplied.

Capture-based aquaculture often involves transportation from the capture location to the place of the farm facilities of live material, including seed or brood-stock. Live material movement for CBA generally involves: (i) large amounts of live material; (ii) long-lasting durations (from days to weeks or months); and (iii) major gaps (for instance, when there is the trading of seed internationally). Excellent practices may not always be used to avoid or limit their spread and additional action may be necessary for the treatment of disease, infection and parasites.

**Figure 3.10**: Lower tropical farming tackles some of aquaculture's environmental issues. Some of the hurdles for the expansion of rabbitfish aquaculture involve the provision of hatchery fingerlings to replace wild fish, the creation of algal-specific diets and an improved understanding of and mitigation of illnesses.

*Source:* https://www.aquaculturealliance.org/advocate/aquaculture-of-pelagic-fish-part-4/

In the context of wild populations, and in the fields of aquaculture, complete guidelines for FAO Technical Guidelines for the Management of Responsible Movement of Living Aquatically Animals (FAO 2007) are provided. A guiding principle and an example of developing good practices, including transfer and quarantine, can be found in codes of practice for the introduction of alien species by the International Sea Exploration Council and the European Internal Fisheries and Aquaculture Consulting Commission (EIFAAC) and FAO Genetic Resource Management Technical Guidance (FAO, 2008b).

Live materials are collected and shifted from one place to another and stored on cultivation sites, which are open to the ecosystem, and can be deposited in wild closely related species in other areas and in wild populations or genetic material (eggs or seeds in most situations but include brood-stocks in some operations). It may be from wild and/or production facilities to cultivate living material. The entry into the stocks of traditional facilities of illnesses, parasites and undesired genetic material provides the possibility to spread disease, parasites and genetic material to hatcheries in or outside custody areas or at the site of the cultivation facility.

The escaping of live material prior to arriving at the farmhouse or at the institution and the insertion in the wild populations or into culture facilities of foreign oxen, seeds, brood-stocks, diseases or parasites should be avoided when handling, transferring and transporting live materials for the CBA. Furthermore:

- Proper procedure should therefore be developed and used to prevent live material, in egg, seed or brood-stock, from escaping to the wild populations outside the catchment area, and to prevent the transfer of pathogens or parasites to the wild.
- Best procedures must be devised and implemented to prevent diseases or parasites from being transferred from the wild live materials to the live materials or stock currently in cultivation facilities (Rico et al.,2017).
- In the case of transfers and the transport of live material, quarantine facilities and practices should be incorporated into the overall CBA development.
- To avoid the escape of live material and the transfer of genetic material to wild populations outside the capture or culture region, the best methods should be established and applied.
- To avoid the introduction of genetic material into live material currently in culture facilities, best practices should be developed and applied.

## *Grow-out and culture issues*

Capture-based aquaculture is cantered on the cultivation of live animals acquired from wild species and must be retained at a biologically sustainable level as a natural resource, while resolving equality issues, if they are included in a multi-sectoral fishing activity for the target species. Good practices are required in culture, because the needless or excessively high mortality in the aquaculture phase of the CBA inventory will lead to extra pressure on wild fish to replace losses throughout the aquaculture processes and lead to losses and waste of living material.

Less than optimal products can be manufactured, for example in low quality products or in fewer quantities than is good for cultivation, and the demand on wild stocks to produce more live material can also increase.

The construction and implementation of appropriate aquaculture conditions, methods, facilities and equipment relevant to several species participating in the CBA can take place under good practice and standards. Aquaculture facilities should provide the highest survival of wild caught live material and/or optimum production. Live CBA traditional activities should apply best practices, with suitable circumstances, procedures, equipment and facilities, so that adequate conditions are ensured for maximum survival and/or for optimum output from live wildlife. For logistic, financial and animal welfare reasons, particularly during weaning and acclimation, regular inspections and controls of the health of live material in CBA systems are important. Some wild-taken live material won't easily accommodate captivity or be weaned upon the supplied food. These organisms appear hungry and/or not healthy. The problem can be addressed by the communication between CBA fishing companies and hatchery supervisors cantered, wherever possible, on earliest stage larvae to be captured or on stage or steps best suited for effective weaning and acclimatization and transferring them to regulated weaning conditions leading to greater survival and good financial viability and animal welfare. For many kinds of wild animals, weaning CBA live materials and brood-stocks to artificial feed is a vital part of aquaculture. Living material from captured aquaculture is used to eating wild food and has not been tamed on synthetic feed.

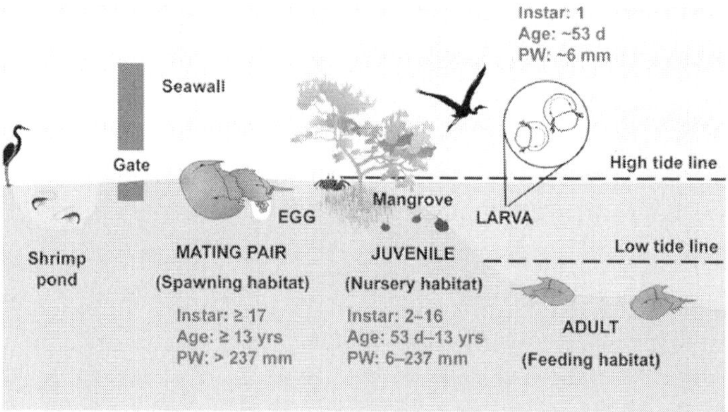


Figure 3.11: The use of hatchery releases with various management targets is referred to as restoration ("to restoring depleted spawning biomass to the level where it again offers a regular substantial yield"), stock improvement ("to increase a natural supply of juveniles and optimize their harvest by eliminating recruitment restrictions").

*Source:* https://www.frontiersin.org/articles/608155

In addition to animal welfare difficulties, weaning can lead to death and injuries and wounded fish may be vectors of disease. It is necessary to identify and remove damaged live materials as soon as possible, and to respond to causes of excessive injury or mortality, in order to prevent the spread of illness and preserve aquatic organisms' welfare. Organisms should be periodically evaluated during the acclimatization and weaning of CBA living materials and wounded, non-adapted individuals should be removed and managed as soon as feasible. There must be every attempt to solve the issues of fish welfare.

### *Issues with feed*

While HBAs and CBAs are both fresh and processed, they tend to be a prominent problem for several CBAs because many are carnivores. HBAs and CBAs use wild aquatic resources as feed. There are still several species that are fed directly on fresh fish derived by capture fisheries (e.g. low-value fishes or invertebrates, known as low value/paste fish), and which do not depend largely on bait fish and forage (note that low value doesn't infer that the species have a potential value for humankind). The broader ramifications of the environment, either indirectly as by-catch or directly as feed-fishing, for removing a big and diverse range of species are frequently ignored. But, considering that many by-catch species are food fish used by others and the vast number of wild fish used to feed, it is impossible to overlook the potential ecosystem implications (Raja & Jithendran, 2015).

The general efficiency in the direct use of fresh, wild-caught aquatic species for the optimal use of biomass as feed for aquaculture due to the poor conversion rate of these organisms as well as the potential health hazards associated with the activity are more generally concerned in traditional operations. There is now an increasing trend to support the use and long-term much cheaper of formulated feeds less dependent on aquatic creatures from fish and/or fish meals. Commercially manufactured formulated feeds for many CBA cultivated species are not generally accessible or currently not commercially desirable.

Many of the complex feeds still rely heavily on wild-fish throughout development. Research is necessary in order to generate cost-efficient artificial feed which ensures minimal mortality during weaning and the excellent meat quality demanded by the market for wild catch aquatic organisms. Research on compound feed is minimal for several cultivated species in developing countries and attention is needed in this field. In addition, the utilization of non-marine (for example, land-based animal by-products, meat meal or blood meal) items as fish-food is also required. In the Hong Kong special administrative Region, fish farmers would prefer to use wild fish rather than pellets of feed in their cultivation since it is cheaper and simpler to get them, for instance, in relation to new feeder types there are often market and socio-economic pressures.

**Figure 3.12**: Aquaculture based on capturing entails catching and then keeping wild animals alive or feeding them to harvest. The purpose is to boost biomass, enhance the quality of raw materials and the chance to act strategically with regard to the market by determining the harvest and sale of the fish. For example, a short and frantic season characterizes fishing for cod. If wild fish are harvested, and kept alive in the cage, fresh cod may not be provided in the short fishing season but throughout the year.

*Source: https://nofima.no/en/nyhet/2013/01/new-book-about-capture-based-aquaculture/*

In the same way that live CBA materials are captured and transported, including feed and broods, it also provides a risk for illness and/or parasites to be introduced into or among wild populations of the same or other species, through the capture, transport and utilization. These concerns are steadily being increased by the increased usage of wild fish for the CBA industry. To address this challenge, treatments to lower or eliminate the burden of pathogens in feed are important. FAO's Health Management technical recommendations on the responsible movement of live aquatic animals are a broad approach to addressing these challenges (FAO, 2007).

The transfer from traditional establishments and from feeding-stuffs of illnesses and parasites into wild populations should be avoided. Specifically:

- Best practices to prevent the spread of viruses and/or pests to the wild outside the catching region of the CBA stock are to be identified and implemented.
- Best practical methods to prevent infections or parasites from wild fish living in the previously cultivated areas from being transferred should be found and implemented.
- These feed organisms in aquaculture should be handled for the reduction in the risk of disease transfer from wildlife caught feed.

In addition to study, a lot of what we know about aquaculture based on catch is based on try and mistakes. Once new vessels expressed their interest in the fishing sector, novice fishermen (and) began with no training, and the same blunders were sadly repeated all too frequently.

Chapter 4

# ADVANTAGES AND WEAKNESSES OF INTENSIVE AQUACULTURE PRODUCTION SYSTEMS

## CONTENTS


4.1 Introduction .......... 90

4.2. Advantages of Intensive Aquaculture Production Systems .......... 90

4.3 Disadvantages of Intensive Aquaculture Systems .......... 104

## 4.1 INTRODUCTION

Intensive aquaculture is characterized by different advantages and disadvantages. These are discussed below as follows.

## 4.2. ADVANTAGES OF INTENSIVE AQUACULTURE PRODUCTION SYSTEMS

### 4.2.1 Large Amounts of Fish Produced

**Figure 4.1**: Fish production has increased rampantly due to intensive fish farming.

*Source: https://animalequality.org/blog/2019/09/30/fishing-industry-destroying-environment/*

With intensive fish production systems, fish products are generally numerous in that they are surplus compared to other aquaculture production systems. This form of intensive agricultural systems has resulted in fish being rendered cheap and a readily available form of meat. Most countries known to be major producers of fish mostly use intensive production systems. The large production of fish is a result of various factors being controlled to ensure the output is maximized. Intensive aquaculture systems utilize artificial conditions that have been well examined and put together to ensure that young fish are able to thrive and reproduce. This form of production makes use of a variety of resources and ensures that they are well utilized. Fish production in such systems is large because fish are well fed. Actually intensive fish culture systems revolve towards fish being well-fed while contained in a given area. The amount of fish produced through intensive aquaculture systems globally is usually high several farmers have adopted this approach (Quesada et al., 2013).

There are several water bodies utilized by farmers allowing them to contribute to the total amount of fish produced through intensive fish production systems. The large numbers of farmers adopting this form of farming have contributed to the large amounts of fish being produced globally. Various technologies are used to ensure that the general harvest is plentiful. These technologies are used in different stages of fish production. These technologies have been used in ensuring that the fertilization process is successfully done resulting in numerous fingerlings being produced. These technologies have also been used in modifying the conditions so that the fertilized eggs are able to survive resulting in a large amount of fish that could be harvested. Usually, the chosen technology is first analyzed before it is used. This is to ensure compatibility with the farming system. Technological compatibility has proven to be an important component in ensuring bounty harvests. These technologies are also used in the harvesting of fish in a manner that they do not tamper with other growing fish ensuring that they grow to maturity.

Some of the technologies used such as the composite fish culture common among most fishing grounds as it has been useful in heavy production of fish from tanks or ponds. When using this technology readily available fish food organism resources are used in each natural niche. In other cases, supplements are used all in a bid to ensure that the harvest is plentiful. The numerous fish produced is because the little of spaces are utilized. For small farmers, they can use tanks that are able to retain water to depths of 1.5-2.0. The minimum water level in tanks is 1.0 meters. The fact that seasonal rivers are also used as resources help in improving the number of harvested fish (Perumal et al., 2015).

This applies to both fresh water and salt water fish. Intensive agriculture led to a significant improvement in the amount of marine fish produced. Intensive marine aquaculture systems have resulted in an increase in the amount of marine fish being exported across the world. It has allowed farmers to manipulate the ecosystem allowing the production of marine fish in unfavorable seasons. The used technologies have enabled climatic conditions to be manipulated to allow them to survive in changing environmental conditions. The technology is also useful in helping generate and use available food resources to ensure that they are well fed and there are suitable conditions for successful fertilization of fish. Generally, the high productivity rates of intensive aquaculture practices have resulted in more fish being put under production.

### 4.2.2 Source of Employment

The growing adoption of intensive fish productive systems has led to an increase in employment opportunities. There are numerous jobs done in this field. However, most employment opportunities may require one to have a significant amount of skill to enable them to work in such farms. Among the job opportunities associated with intensive fish production is the creation of cages. There are farms that use cage systems in rearing their fish. This presents an opportunity for individuals skilled in the art of cage making as they could make them and sell them to farmers. With small-scale farmers adopting this form of aquaculture, there is a growing demand for cage makers. Another area of employment is the handling technologies used in intensive farming. There is a variety of technology used in intensive fish production systems that requires skilled individuals to handle and monitor these technologies. This is an avenue for employment as individuals will be involved in routine maintenance practices of such technologies. Others could be involved in using these technologies in controlling variables to ensure the high productivity of farms.

**Figure 4.2**: Many employment avenues have been created through intensive fish production systems.

*Source: https://orissadiary.com/revolution-fish-farming-women-self-help-groups-odisha-shows-way/*

In both large and small farms can be used in the feeding fish. These individuals are given the types of feeds to be given to fish and the time these feeds should be administered. There is a high demand for such individuals more so in developing countries as feeding technologies have not been well embraced requiring high amounts of labor in feeding fish. There are employment opportunities for those involved in the harvesting fish. Though technology can be used in the harvesting fish, there is still a need for human supervision to ensure that young fish are not captured or killed in the harvesting process. There is the area of sorting fish according to their species and size before they are taken to markets for sale. Rather than working for someone else, it may be advisable for one to start their own farms and rear fish for sale. As long as one has sufficient funds, they are able to start their own farm and employ other individuals to work for them.

**Figure 4.3**: Fish markets have grown due to the production of more fish.

*Source: https://www.newindianexpress.com/states/odisha/2019/jun/15/hygienic-fish-markets-abandoned-in-odisha-1990414.html*

Another area of employment is in the preparation of fish for markets. The growing trend of intensive aquaculture farming has resulted in several fish industries being established to process fish for both internal and external markets. The establishment of such industries has resulted in several employment opportunities being created. The growing production of fish has resulted in growing fish markets and therefore more employment avenues for individuals. Intensive aquaculture production systems form a significant percentage of the job opportunities created in fish production. Another employment opportunity is as an intensive aquaculture production system consultant (Anderson et al., 2019).

### 4.2.3 Increased land productivity

Intensive aquaculture production systems have resulted in wasted land being put to good use. This increases land productivity. This form of production system is founded on maximum use of available resources among them land. As the production system is not fixed to water sources, wasted land can also be used in improving land productivity. This is achieved by the construction of ponds on such lands. However, there are certain wastelands that cannot be used. They include those that have been contaminated with radioactive or toxic wastes. The method champions maximized use of space.

**Figure 4.4**: Wasted land are made useful through the construction of ponds.

*Source: https://agro4africa.com/fish-pond-design-and-construction/*

In most cases, lands having lots of clay soil have not been heavily used. Fish farmers take advantage of this fact and establish ponds in those areas. Such soils are more advantageous as they save farmers costs due to the construction of cemented dams. Clay soils are good water retainers and can hold water for longer durations of time. Farmers take advantage of land found near lakes and rivers as it keeps them in close proximity to a water source. Improving productivity of land also includes using fish farms that were once in use and are no longer active. This includes derelict and swampy ponds being renovated to facilitate fish culturing. This is achieved through desilting, dewatering, repairs on embankments and the creation of an inlet and outlet. For owners who do not know what they can do with their lands, they can opt to lease them to farmers who will pay them a certain amount for a given duration of time (Pandey et al., 2014). The payment methods and everything pertaining to the leasing of land should be handled by the farmer and the owner. In some cases, a third party is involved during the process to make the formalities more official. For small scale farmers, they can use a small portion of their land in establishing the ponds. They can use simple materials that are readily available such as polythene paper to help them manage water retention.

In some countries, the government allocates government land to be used in intensive aquaculture production. In some cases, the land given is found next to the water source while in other cases the land is given and farmers have to cater for watering systems. When constructing ponds, farmers need to take into consideration some of the materials needed for ponds to be established which include: already existing ponds being desilted, shallow ponds being deepened, marginal areas of water bodies being impounded, embankments being constructed, inlets and outlets being established and the establishment of other civil structures as per the project requirement. For those who have direct access to large water bodies, they can use cages to conduct intensive fish farming though the challenge may be ensuring the safety of fish.

### 4.2.4 Source of food

Fish gotten from intensive has become a major source of food. It has been useful in meeting the growing global demand for fish for a number of years to come. Intensive fish farming is known to contribute to the large percentage of fish produced globally with some countries being known to be major producers of both fresh and marine fish. With the ever-growing increase in population, there is a need for the production of more food to meet the demand. This is achieved through intensive fish production systems. This has resulted in many saying that this sector for food production is eventually growing. Intensive fish farming has resulted in aquaculture being identified as a solution for the food insecurity issue and also improve nutrition among different parts of the country.

**Figure 4.5**: Fish is considered a healthy source of food.

*Source: https://www.healthline.com/nutrition/12-omega-3-rich-foods*

Intensive fish production systems have resulted in the development exhibited in aquaculture by increasing fish production. This will enable the fish industry to meet the food demand. This makes fish an alternative to food sources. These production systems have become a major source of food more so in developing countries. Farmers who practice intensive fish production at a small-scale sometimes do the production for subsistence purposes. This means that fish is reared be used at home. This makes farmers from large families to be able to meet the food demand. Globally, there is a lot of food insecurity and there are major shifts from initially depending on plant products. However, researchers looked at other sources of food and fish was among the food sources. Countries such as Japan took advantage of this knowledge and decided to heavily invest in intensive fish production systems. This has resulted in them being referred to as major fish producers. Some of the fish produced are consumed locally consumed while most of it is exported to other countries (Panigrahi& Azad, 2007).

As a source of food, fish obtained from these production systems also have nutritional benefits. According to research done by the FAO about one billion people globally rely on fish as their major source of protein. Globally, the annual average fish consumption is about 16.1 kg per capita. In several countries globally, fish are produced mainly for consumption. This is well demonstrated in countries such as Vietnam and Thailand. As fish is becoming a more common source of food for individuals, its consumption rate has increased. As mentioned earlier, the increase is due to intensive fish production systems that make use of technology in conducting these activities. Intensive fish production systems produce large amounts of fish that have made fish become a much cheaper source of food. For families living in poverty or those from low economic backgrounds are able to access this source of protein which will be useful in ensuring that they are healthy. This is because fish is a much cheaper source of food and is affordable for them. They also benefit from the fact that fish usually has high protein levels compared to other sources of food. It is also rich in vitamins and minerals. Therefore intensive fish production systems are a useful method of ensuring that families from poor backgrounds are able to eat healthy meals by ensuring that fish is readily available for all and they can get a variety of meals.

**Figure 4.6**: Fish are a good source of omega and very nutritious.

*Source: https://www.usaid.gov/results-data/success-stories/fish-farming-brings-food-and-nutrition-security-sierra-leone*

By selling fish obtained from intensive fish production systems it becomes an indirect source of food, therefore, ensuring food security. Also, intensive fish production systems are a major source of income for most families allowing them to purchase food. Intensive fish production systems in most developing countries have become a major source of livelihood. In developing countries, these production systems have helped improve income resulting in increased food purchasing power as well as diversification (Ottinger et al., 2016).

### 4.2.5 Use of recycling systems

One of the advantages of intensive fish production systems is that it revolves around the use of integrated recycling systems. This is really crucial when aerators have to be used during the process. The challenge with the different types of fish production systems is that it uses large amounts of water. This results in large amounts of water being wasted and this poses a threat to the environment. This is commonly experienced when pisciculture is used as it uses about a million gallons of water for a particular acre of land in a year. Most farmers practicing intensive fish production have opted to use extend water purification systems. These systems allow them to resize local water. This makes the process more efficient and productive as it makes use of water sources that could have been left to waste.

**Figure 4.7**: Water recycling systems are used in efficient water management.

*Source: https://en.wikipedia.org/wiki/Recirculating_aquaculture_system*

With intensive fish production systems, large-scale farms are known to make use of a recycling system developed by the New Alchemy Institute.

The same institute also refined the system which makes it more efficient. This system involves large plastic fish tanks being established in greenhouses. Once put in the greenhouse, a hydroponic bed is placed closed to the plastic fish tank. The type of system requires the farmer to choose the appropriate species of fish. In most cases, those who use this system opt to use tilapia being reared in those tanks. The benefit of rearing tilapia is that it can eat algae known to grow on the tanks. Water from the tank is circulated in the hydroponic bed. This allows farmers to use wastes gotten from tilapia used in fertilizing crops. The hydroponic bed has properly cultured microorganisms that have been useful in converting ammonia to nitrates. The phosphates and nitrates are used in fertilizing plants. There is a hydroponic medium that allows wastes to be strained which resulted in aerated pebble-bed filters.

When this recycling system is properly manipulated, the outcome is the production of large edible protein in a given unit area. When hydroponic beds are used, farmers need to know that different plants can be used in such beds. Most farmers opt to grow herbs are they require premium prices for little amounts throughout the year. Most of the herbs from hydroponic bed are sold to restaurant wholesalers. Most of the recycling systems are used and live in green houses. This means that they are able to adapt to different climates more so tropical climates. After the purification process, farmers should ensure that water used to rear fish is slated to maintain the electrolyte equilibrium required by fish. With the recycling systems provided by intensive fish production systems, farmers can use a variety of tools to ensure that the fish are healthy without them having to deal with the hustle of water. In some cases, the relevant veterinary authority has advised that ultraviolet ozone disinfectant systems can be used in properly managing recirculated water, therefore, ensuring that tilapia are healthy. Among the projects on recycling systems deals with integrated recycling systems being combined with urban farming. This project has been initiated in Sweden under the Green-fish initiative (Opiyo et al., 2018).

It is also a good way of ensuring proper water management. This ensures that available water resources are well utilized therefore allowing these production systems to run for long durations of time. The presence of water purification systems allows farmers using this production system to use available water from a variety of sources and purify it for use in rearing fish. The other added advantage is that water released through the outlet can be collected and taken through the purification process allowing it to be reused in rearing fish. This allows gallons of water from being misused. It also saves farmers from spending more money for buying water. It also ensures that there is a continuous water supply even during dry seasons or when there are water shortages more so in urban areas. Intensive fish production systems ensure that every drop of water counts.

### 4.2.6 Source of income and revenue

Like other fish production systems, intensive fish production systems have become a source of income for most families. Intensive fish production systems have created numerous job avenues for individuals allowing them to earn money allowing them to meet their family needs. There are those who get income from managing the production systems. Others get income by selling technologies used in the systems while others gain income from selling fish to local farmers or exporting them to international fish markets. Countries will several intensive fish production systems are known to earn a lot of revenue. When the systems run, the government gets revenue through paying taxes. The same applies when fish is sold to both local and international markets. This revenue is used in developing the country.

**Figure 4.8**: Government earn revenue through the exportation of fish.

*Source: https://www.referenceforbusiness.com/encyclopedia/Gov-Inc/Income-and-Revenue.html*

This is evident in countries like Japan. The country earns a lot of revenue from the sale of fish. They are known to be a global producer and seller of fish. In Japan, fish farmers conduct intensive fish farming which results in them having surplus produce that is enough for subsistence use and also exportation. The funds have been used in making the needed development in the country. The revenue has been used in establishing schools, roads and hospitals (Otta& Hoque, 2020). The revenue has also been used in improving the standards of living for those in Japan. The same case is exhibited in countries that carry out intensive fish farming. This has resulted in many foreign investors more so those looking for business opportunities in developing countries invest in intensive fish farms. Also, countries practising intensive fish farming are known to attract a lot of investors to generate revenue for the country.

### 4.2.7 Fosters research and technology use

Intensive fish production fish farming systems are more advanced compared to extensive and semi-intensive farming as more research and technology goes into it. The technologies used in intensive fish are carefully used with the aim of improving accuracy, repeatability and precision during fishing operations. It is also useful in creating autonomous and continuous animal surveillance and providing more information to allow relevant stakeholders to make sound decisions. It also reduces the risk presented to farmers such as drowning in water. It also reduces the heavy dependence on manual labor. By so doing the welfare and health of the fish will improve. The same case applies to its productivity.

Technologies used are for both sea and fresh fish farms. Among the technologies used in the Atlantic Sapphire Blue house schematic is the Recirculating Aquaculture System. This technology is mostly used in intensive fish farms that use recycled water and their fish farms are interconnected with plant farms so that water from the fish farms is used in watering plants. This technology is used in reducing carbon footprints of seafood. The advantage of rearing fish in these farms is that they are able to adapt to traceable and controlled environments with the exception of antibiotics and hormones. These technological systems can be used in every urban setting. A test facility in Denmark has been made the largest facility offering this technology. In this facility, fish species such as salmon are able to thrive as the conditions presented are ideal for their growth. This technology has attracted many investors and has resulted in more land being given for fish production. Among the companies that use this technology is the Nordic Aquafarm. The farm is expected to produce about 33,000 tons of salmon.

**Figure 4.9**: RAS is among the technologies used in intensive fish production systems.

*Source: http://aquaculture-israel.com/Technology/re-circulation.html*

Among the focused areas for technological innovation in intensive fish farming is the Nordic Aquafarm's Belfast plant. The plant is known to deal with fish species such as salmon. In countries like the US, these fish species are domestically produced. However, further research has opened doors for other fish species to be reared under the technology more so with RAS developers. Farmers are adopting the use of the RAS and creating more demand for salmon. The use of this technology has expanded so much that countries like Vietnam have expanded their fish production so much that the amount of fish harvested is sufficient to allow its distribution to various areas (Amenyogbe et al., 2018). The RAS technology is mostly used in onshore fishing systems. Further research has been done so that more innovation can be done to complete the already existing fishing systems. This has resulted in the production of genetically modified salmon. There are other companies that develop and sell their technologies to intensive fish farmers. Among the technologies is the Aqua-bounty Technologies. The company has been selling its products in Canada. Several modifications are being made to the genetic modification of salmons. This has been achieved through genetic engineering. Genetic modification of salmon has resulted in its reducing maturity rate.

Research and technology have been used in the field of farm management systems. In this field, it is important to ensure that most of the innovations done revolve around improving infrastructure. Most of the technologies used include sensors and other equipment that allows farmers to get data needed in ensuring farm efficiency. The data is also used in improving conditions so that fish are able to thrive in the given conditions.

The Innovation Network Corporation of Japan is known for its technologies and their success rates. This has led to countries like Singapore having companies get about 8.2 million seed rounds from the Japan Corporation. Singapore is working towards the establishment of an Aquaculture management system. This system is referred to as Umi-Garden. The Aquaculture management system gives farmers insight of management software and fish behavior analysis by the use of sensors. Another company known as Aqua-byte is also working towards developing machine vision hardware and well as a software tool to be used in conducting data analysis. The company is working towards the elimination of sensor systems in intensive fish production systems. To replace it will be subscription-based models.

Another area where these technologies are used is the feeding of fish. With intensive fish production systems, much emphasis is placed on availing feeds for fish. By ensuring that they are well-fed, the farmer will be assured of high productivity rates. Farmers incur high costs on fish feeds. In Aquaculture, fish feeds are made up of a mixture of proteins, oils, carbohydrates, vitamins and other ingredients. Feeds given to animals vary depending on the species of fish. With technology, the feeding of fish becomes much easier. Technology has mechanized in a manner that allows remote feeding of fish. It released controlled feed amounts (Ngajilo&Jeebhay, 2019).

The other advantage of using technology is that fact that it allows genetic modification of fish genes allowing the production of fish with desired genes. For instance, genetic modification could lead to fish having a faster maturity rate and growing to have big sizes compare to the normal expected size. The genetic modification also facilitates the production of fish species that easily breed and have high chances of fertilization resulting in the production of more fish. It has also opened up the ideology of other fish species being reared through intensive fish production systems.

## 4.3 DISADVANTAGES OF INTENSIVE AQUACULTURE SYSTEMS

### 4.3.1 High disease incidence

With intensive fish production systems, there is a high likelihood of fish getting diseases. If one fish gets sick, there is a high likelihood that other fish will get sick more so with fish reared in cages. The diseases spread fast to other areas. Though it is maybe quite easy to spot and treat the diseases once identified, there is a high likelihood that once the disease begins to spread there will be severe damages. Disease incidence can be experienced in both fresh and salty water production systems. Some of the diseases are specific to certain fish species while other diseases affect all fish. One of the diseases affecting freshwater fish is piscine tuberculosis that is caused by a bacterium known as *Mycobacterium marinum*. This disease affects all fish and can be identified through weight loss, papules, bulging eyes and loss of appetite. The treatment method in this case is amputation.

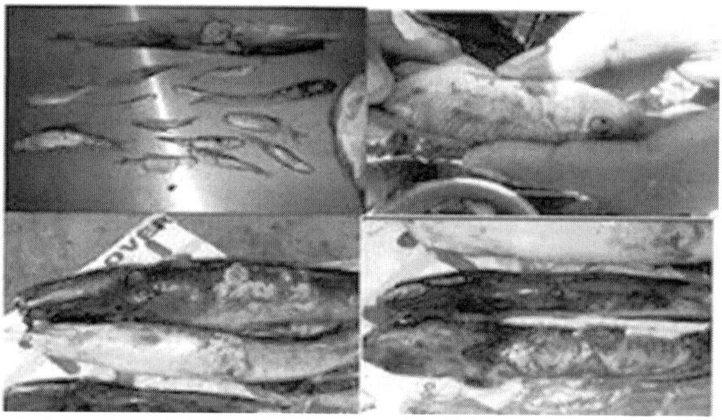

**Figure 4.10**: There are a variety of diseases that could affect fish in intensive fish farms.

*Source: https://www.sadc.int/news-events/news/press-release-outbreak-epizootic-ulcerative-syndrome-eus-disease-sadc-region/*

Another disease common in fish is Fin Rot and it is also caused by bacteria. It affects all fish species and is evident through the presence of erosions on fin edges. The treatment method in this case will be to improve the water quality, feeds given to fish should have high levels of vitamin C and the removal of fin-nappers. Swim bladder disease is a disease caused by digestive issues and bacteria. This disease can affect all fish species and causes them to lose the ability to float in water. In this case, farmers will be required to regulate water temperatures. In intensive fish farming, water is regulated through temperature systems used in the farm (Narbonne et al., 2021). Other bacterial diseases that affect all fish species more so those in cages are Vibrosis and Pseudomoniasis. In the case of Vibrosis, it causes discoloration, bulging eyes, staining of fins and lethargic behavior while in Pseudomoniasis, the fish suffers from hemorrhages in the mouth. Both diseases are treated through the administration of medicated food.

With fish diseases, there are those caused by viruses. With viral diseases, they affect all fish species and farmers may be required to take the necessary precaution to ensure that the likelihood of such diseases occurring is minimal. An example of such disease is lymphocystis and is detected when fish are noted to have cauliflower-like growths around the eye region. To best prevent the disease, it is advisable for the farmer to administer vaccines. There are other diseases caused by a fungal infection. There are those that affect full-grown fish while others affect fish while they are developing eggs. Examples of such diseases include Egg fungus and Saprolegnia fungus. In the case of Egg fungus, the disease affects eggs and is detected by fungal growths on the eggs. The treatment used in this case is using methylene blue in medicating the hatching tanks. Affected eggs should be removed to reduce the likelihood of diseases spreading. In the case of Saprolegnia fungus, it is detected by the presence of whitish fur-like growths. The method of treatment in this case is giving fish food rich in vitamin C or using a water bath that has been medicated (Muhala et al., 2021).

For intensive fish production systems rearing marine fish species, there are certain diseases that could affect different fish species. Some of the marine fish diseases include Marine velvet and is caused by a parasite known as Amyloodinium. Some of its symptoms include fish having a powdered appearance, disorganized and gasping swimming. Another disease is the Velvet disease caused by a number of parasites among them being Oodinium. It is noted through the presence of golden dots on fish. Another common symptom is the tendency of fish to rub their skins on rocks as they swim. Egtyed virus is among the viral diseases that affect saltwater fish. A common symptom is internal and external hemorrhaging of fish. Some of the diseases affect both fresh and salt water fish. With the high likelihood of diseases occurring, farmers practicing intensive fish production are likely to incur a lot of costs treating the diseases. In most cases, farmers opt to rear fish species that are less vulnerable to diseases. This means that the fish species have good genes.

To avoid a high incidence of diseases, there are cases where farmers rear genetically modified fish. This is achieved by research checking on the type of genes that can be dominant over diseases. Samples of fish are taken to laboratories where they undergo genetic modification and are tested to check whether they can be easily affected by bacteria and viruses. If they prevail, they are used in large-scale production of fish. The other challenge with intensive fish farms is that farmers have to incur high costs in purchasing pesticides and other chemical treatments to kill bacterial and fungal pathogens. These chemicals have to be frequently used to avoid the chances of these pathogens building up in the water.

This places high emphasis on the need for water purification systems to avoid contaminants getting into ponds. The danger of high disease incidence is that there is a high likelihood of farmers incurring losses due to the death of fish. This will result in farmers incurring losses and also reduce the amount of fish that could have been harvested during that season. This further contributes to the problem of food insecurity experienced globally. It will also increase the funds that could be used in a production season.

### 4.3.2 Environmental degradation

In as much as intensive fish farming was taken as an approach to improve fish production and increase food security. It has both its positive and negative effects on the environment. One of the major negative effects of intensive fish farming is environmental degradation. Some of the activities done in intensive fish farming are negatively affecting natural ecosystems by reducing oxygen concentrations among others. Most of the environmental problems were noted after the production systems were fully integrated (Mohapatra et al., 2013). Some of the negative environmental effects were captured by the press and shared with the public who have also raised their own concerns. One major part of intensive fish farming systems is the feeding of fish. The feeding process is intensively done with fish being given a variety of feeds having different components. As feeding is intensively done, it results in the buildup of nutrients and effluents in water. This has resulted in a lot of environmental degradation more so in water bodies used in intensive fish production systems.

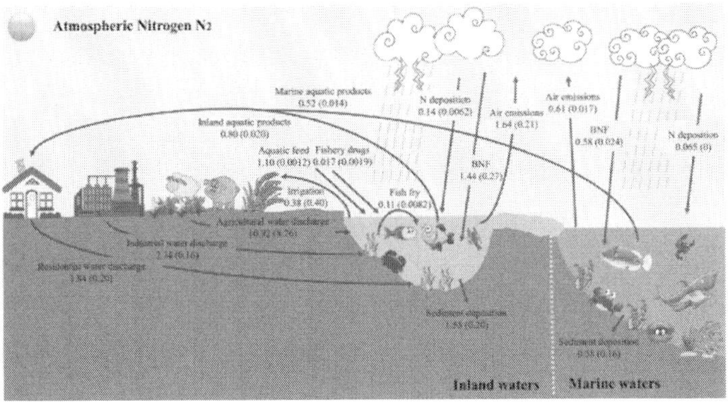


**Figure 4.11**: Some of the activities done in intensive fish farms have negative effects on the environment.

*Source: https://www.nature.com/articles/s41598-018-29214-y*

Some of the nutrients that remain in water due to the excessive food given to fish include ammonia and ammonium which are the major forms of nitrogen in the water. These two nutrients in high amounts can be very toxic to fish. These two nutrients are very soluble in water. In water that has high oxygen concentration levels, ammonia is transformed into nitrate. Another nutrient commonly found in water is nitrate. This nutrient is mostly found in lakes and streams. As it is very soluble in water it results in reduce oxygen concentrations in water. In large amounts when water filled with nitrates is consumed, it results in several health problems. With the heavy presence of nutrients in the water, a process known as eutrophication takes place. This process is caused by algae feeding on some of the nutrients in water, therefore, breaking it down. Some of the nutrients block pipes and cause the formation of scum on water surfaces (Mishra et al., 2017). The presence of these nutrients results in the increase of algae on water surfaces and when they die and decay, they result in water having a bad taste and odor. The decay process also results in water having reduced oxygen concentrations. This results in the death of large amounts of fish. The large amounts of nutrients in water result in the generation of harmful algal blooms caused by nutrient enrichment. These bacteria result in some of the diseases exhibited in fish. This increases the incidence of fish suffering from both fungal and bacterial diseases.

The heavy presence of nutrients makes it hard for farmers to control oxygen concentrations in water. This means that they should look for other ways of ensuring oxygen concentrations are maintained in water. This includes water undergoing a purification process or a re-oxygenation process. This includes farmers growing plants like bamboo near water bodies. Plants like bamboo are known to naturally purify water. Recycling water is also another way of reducing the high nutrient concentration in water. This is achieved when water is used in rearing fish being released from an outlet where it is collected and used in watering plants. When watered, plants absorb the nutrients. It also increases soil nutrient levels.

Another way in which intensive fish farming causes environmental degradation is through the use of chemicals. There are numerous chemicals used during the process of rearing fish. There are chemicals used in preventing the incidence of parasites, fungi and bacteria. In most cases, these chemicals are sprayed over water. If the individual spraying it does not spray it properly then the chemical will go to adjacent lands and cause leeching among other effects. Some of these chemicals tend to be poisonous when they come into contact with human beings and animals. Some of these chemicals also cause air pollution. The chemicals used also affect other aquatic animals resulting in their death (Michael et al., 2014).

The negative effects of intensive fish farming with relation to chemical use are well known among people. In such farms, several fish are kept therefore large amounts of food are required. This means that the nutrient build-up will increase to a high level. This is increase with the fact that fish produce their own wastes. This increases the nutrient build-up in the water and its surrounding environment. This has led to significant decline in oxygen levels. It has also caused the creation of dead zones and algal blooms. Other than chemical use, there are concerns about the use of antibiotics in preventing diseases. There are concerns that antibiotics will have their own effects on the surrounding environment and other aquatic animals. There are more concerns on the issue of nonnative fish could escape and results in wild fish competing with other fish for food. This could result in native fish being displaced in the natural ecosystems. However, though there are challenges with its effects on the environment the relevant stakeholders are actively working to tackle the issues presented by chemicals and medicines in the environment.

There is significant progress made in the intensive fish farming industry. This is beneficial in ensuring the well-being of different ecosystems. This is useful in ensuring food stability. There are several regulatory bodies that have been involved in making the necessary improvements so as to ensure that there are minimal effects of nutrients and effluents due to the production systems. These bodies are also involved in ensuring that all the preventive approaches have been adopted. Among the measures taken to reduce the negative effects is to ensure that locations of such fish farms are properly selected. It is advisable that farmers choose a location where there are strong currents that can disperse the effluent so that the effluents do not concentrate in one area (Marques et al., 2020). Also, it is advisable for farmers to shift their farms from one place to another so that they can reduce the incidence of effluents building up in one area. One of the strategies that have been useful in reducing the negative effects of effluents in water is the use of land-based aquaculture. Through the regulation bodies, the use of antibiotics has also reduced with also the vaccinations administered to fish are more safe and effective. To deal with the issue of fish escaping from farms, most large-scale farms integrated the use of underwater cameras that will closely monitor cages and allow farmers to inspect cages and check whether there are escape spaces for fish.

### 4.3.3 Unhealthy products in fish

With intensive fish production systems, fish are fed on feeds that have many chemicals in them. Chemicals in the feed are used in increasing the maturity rates in fish. These chemicals are ingested by fish and they become part of their body. This means that as human beings and animals eat these fish, they consume the chemicals in them. Chemicals also find their way to fish bodies through the use of pesticides in controlling pests' incidence in water. The same applies when bacteria and fungi have to be controlled. When antibiotics are administered to fish, they find their way to the bodies of human beings. These chemicals also reside in water after some of the fish die and decompose releasing some of the chemicals in the water. The production of fish filled with chemicals is unhealthy for animals in different ecosystems. For instance, when big fish such as sharks get to feed on such fish then chemical buildup will be exhibited in their bodies. This will result in populations of such fish reducing as these chemicals alter their genetic modification that may render the animal unable to breed or fertilize their eggs. This results in a reduction in the fish population. In some cases, if the chemicals are very toxic then it results in the death of these animals.

For human beings, the consumption of fish containing these chemicals may result in health complications. Some of these chemicals may alter the normal functioning of the body in that fertility rates, metabolism and respiration rates reduce. Some of the chemicals used contain carcinogenic materials causing people to suffer from cancer among other diseases. In some cases, these materials result in many bodily alterations causing children to be born with defects. With exposure to pesticides, humans are likely to suffer from skin allergy, congenital diseases or physical deformities. With the exceeding presence of chemicals in foods, human beings are becoming more aware of what they are consuming. Many are shifting towards the consumption of foods that do not contain high amounts of chemicals. This means that markets for such fish will reduce (Lulijwa et al., 2020).

### 4.3.4 Overfishing

With the growing demand for fish across the world, intensive fish farming has been adopted in different countries across the world to meet the demand for fish. Intensive fish production systems cause the production of several fish which can be harvested and sold in fish markets across the world. In as much as this is beneficial, there is the challenge presented. The fact that farmers can harvest large amounts of fish will encourage bad fishing habits such as overfishing so as to meet the demand for fish. This is very dangerous in that if the amount of fish harvested in a certain season does not meet the demand, some fishermen may harvest more fish from natural ecosystems. This means that premature fish and natural coral reefs will be harvested. This destroys natural fish ecosystems. This will tamper the normal fish production resulting in reduced fish populations which may result in farmers losing their jobs in the future.

### 4.3.5 It is expensive

Intensive fish production systems are different from other production systems as all activities are modified to ensure that surplus amounts of fish are produced. There are several activities that make the process more expensive. For one, more workers are needed in monitoring, feeding and harvesting fish. This means that the farmer will have to incur higher costs in paying employees compared to other fish production systems. To reduce the labor requirement, farmers have opted to use technology in conducting most of the farm practices. This includes using technology in feeding fish. Some of the technologies used are useful in getting data on various variables used in making decisions. The technology is also used in conducting the research needed in the field. Underwater cameras are used in checking whether fish are escaping from their cages. In as much as these technologies are useful in preventing accidents associated with fish farming, they are very expensive. The farmer will have to incur large amounts of money in purchasing and maintaining them.

Farmers will also have to incur high costs when purchasing cages used in rearing fish. Fish cages cost a large sum of money purchasing and maintaining them. Large-scale farmers have to invest large sums of money in purchasing durable cages (Luis et al., 2019). High costs are also incurred when farmers have to purchase water treatment systems. This is when the farmers are using recycling systems. They have to purchase purification systems which are quite costly. Farmers have to incur high costs in purchasing feeds and pesticides. With intensive fish farms, lots of the emphasis is placed on fish getting enough amounts of food to ensure fast growth and also ensure that they are healthy.

**Figure 4.12**: Variety of feeds can be used in rearing fish.

*Source: http://www.ep-machine.com/Fish-Feed-Extruder/animal-feed-puffing-machine-40.html*

### 4.3.6 Requires skills

With intensive fish farming, farmers are required to hire highly skilled individuals who will be able to keep up with all activities during the rearing process. These individuals need to have knowledge on how to manage the different technologies used in rearing fish. Usually hiring such employees will be more expensive as most of these individuals are usually hard to find as they are few. In as much as the individuals possess the required skills they need to know how to apply them in the various areas needed.

Chapter 5

# DEVELOPMENTS IN INTENSIVE AQUACULTURE

## CONTENTS


5.1 Introduction .......... 114

5.2 Why Intensive Aquaculture .......... 118

5.3 Difficulties of Intensive Aquaculture .......... 120

5.4 Commercial Aquaculture Production Technology .......... 122

5.5 Development of Controlled Intensive Aquaculture with Limited Water Exchange and Adjusted Carbon to Nitrogen Ratio .......... 130

5.6 Development of Sludge Remover for Intensive Aquaculture .......... 131

# 5.1 INTRODUCTION

Intensive aquaculture is a type of fish farming where the fish are stocked at an astonishingly high rate so as to obtain good amounts of feed from their ecosystem. In this type, the fish are quite dependent on the feed that has been artificially provided to them where water is replaced or replenished at very high rates so that it can maintain the amounts of oxygen to be enough and to also remove the waste. With all this, the rate of water replacement and feeding will determine the amount of fish or the stocking rate per unit area that can be supported (Liao et al., 2009).

**Figure 5.1**: *An intensive aquaculture system for raising tilapia.*

*Source; aquaculture production technology*

This type of aquaculture requires good infrastructural facilities, large amounts of capital and a person with good managerial skills. They fully depend on commercially produced feed, a system that will supply oxygen to the water used by the fish and good circulation of water. Most of the individuals that are involved in this type of system are farmers that are producing fish for the commercial scale, the corporate sector and entrepreneurs that have fully invested in fish farming. With this system, they have been seen to help the communities that live around the fish farms i.e. it provides employment to individuals in the nearby community and lead to the development of ancillary industries.

For this system to fully work there are requirements that are needed for this system to fully work. These include; seed feed (these are feeds that are needed to feed the fish from when they are purchased as fingerlings or hatched in the hatchery to when they are ready for the market, the feed is of different sizes and nutrient concentration based on the type of fish and the age of the fish), lime (also referred to as dolomite pesticide the farmers use lime or commonly referred to as calcium hydroxide or hydrated lime whose main purpose is an increase the hardness of water which is done by the calcium component of line while the hydroxide component increases the alkalinity of the pond (Li et al., 2011). Many farmers use it to primarily raise the pH of the mud below the pond and water (at an evidently faster rate), energy (good supply of electricity due to the various machines being used in the facility i.e. the machines for water circulation and water supply, most of the time renewable energy is recommended because it might be cheap o the farmer, most farmers will prefer solar energy), oxygen supply (this involves the farmer using oxygen pumps to pump oxygen to the pond using pipes, most of the farmers bubble oxygen from below the pond) and high-level management (the farmer or cooperate company needs to have good management so as the manager should have good skills and adequate knowledge about the fish and they are supposed to be managed in order to make profit from them.)

When compared to extensive fish farming the cost of the inputs per nits of the fish weight is far much higher and this is brought about because of the cost of the fish feeds. The cost of feed of fish is much higher because it contains a much higher protein content i.e. the protein content is 6% much higher than protein found in cattle feed and there has to be well-balanced composition of amino acids. The high protein requirement is brought about by the fact that aquatic animals have a higher feed efficiency i.e. they have a high Feed Conversion Ratio (FCR), which is the kilograms of feed per the kilograms of the animal that is produced. For fish such as salmon, they have an FCR of around 1.1 kg.

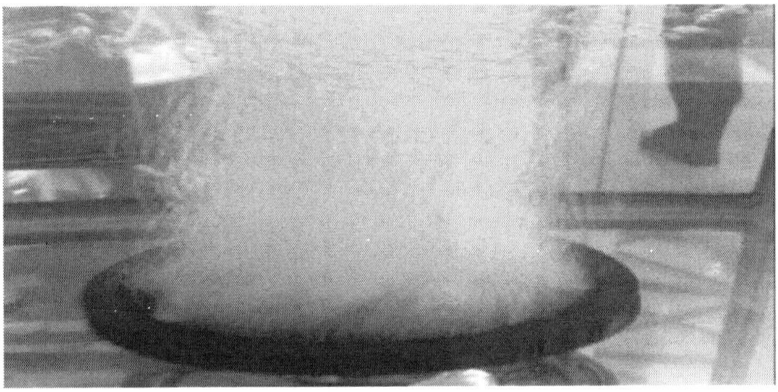

**Figure 5.2**: An oxygen diffuser used in the addition of oxygen to water.

*Source; Wuxi Qiuyuan environmental engineering*

The aeration aspect is very essential because fish need oxygen found in water for them to survive. Most farmers use bubbling (farmers can use two options i.e. subsurface bubbling or bottom up bubbling, surface aeration will involve agitating the top surface of the water in order to mix oxygen and the top layer of water, with this system the fountains will pump water into the air where it will create small droplets which will collect oxygen in the air and fall back into the water in the pond, in the bottom up approach, oxygen that has been compressed is pumped into the water through diffusers that have been placed at the bottom of the pond) or cascade flow that involves water flowing directly into the pond and out through the overflow to allow for water rich in oxygen to enter the pond through one end and exit through the other end) and use of aqueous oxygen (Labh & Shakya, 2014). Some fish such as *Clarias* spp. can directly breathe atmospheric oxygen and they are able to tolerate waters that have low levels of oxygen as well as tolerate high levels of pollutants as compared to salmon or trout, which cannot tolerate as much. This makes the *Clarias* spp. a good species for intensive fish farming because water purification and aeration are less necessary for their survival.

**Figure 5.3**: *Clarias* spp is one fish that is known to be used for intensive fish farming.

In this system, the amounts of waste produced are quite high that means there is an unnaturally high amount of organic matter produced and released into the environment. The wastes may be in form of particulate and/or in soluble , which can include fish feed that they have not eaten and other excretory products. There is a high release of nitrogen and phosphorus in particular and they are released in solute form into the water. Hypernutrification is a condition whereby there is a release of nutrients into an aquatic environment or ecosystem and eutrophication that is the process whereby an entire aquatic ecosystem or a part of the aquatic environment becomes enriched with nutrients and minerals e.g. phosphorus, nitrogen and other plant nutrients are the two main processes that discharge wastes into the environment. Hypernutrification is substantial and it is the gradual increase in the concentration of dissolved substances in the water that comes mostly from compounds of nitrogen i.e. nitrate and ammonium and phosphate. With the water having accumulated the nutrients, it leads to the accelerated growth of algae and other high forms of plants, which then leads to undesirable disturbances to the quality of water and the balance of organisms.

Apart from intensive aquaculture, there is the extensive one and the semi-intensive one. When fish farming involves extensive fish farming, the fish tend to feed entirely from the food web. This happens within the pond and can be supplemented by the addition of fertilizers and manure to help in the growth of water plantations. The ponds that are used may either be natural ones or artificial and the plants that are normally promoted include zooplankton (mostly small crustaceans), aquatic vegetation and phytoplankton (microscopic plants) that help in finishing the cycle of the aquatic food pyramid. As a matter of fact, with this type of fish farming, there is an improvement in the marketable animals that belong to the natural ecosystem. Common carp among other fish species are known to be bred using the extensive method in the European Union (Kalbassi et al., 2013). This is however an area where intensive fish farming is being practiced widely. In Asia and Africa, there is the extensive fish farming of tilapia and carp. There is one method that China uses involving the bighead carp. It involves growing them in small lakes and reservoirs but in this case, they use neither the fertilizers nor the feeds. These bighead carp are normally stocked with other fish. These fish are also grown in polyculture ponds and pens. In this case, there is normally the use of organic fertilizer, which helps in the growth of organic foods. This is majorly done predominantly when the bighead and the silver carp are the major species being reared in the pond.

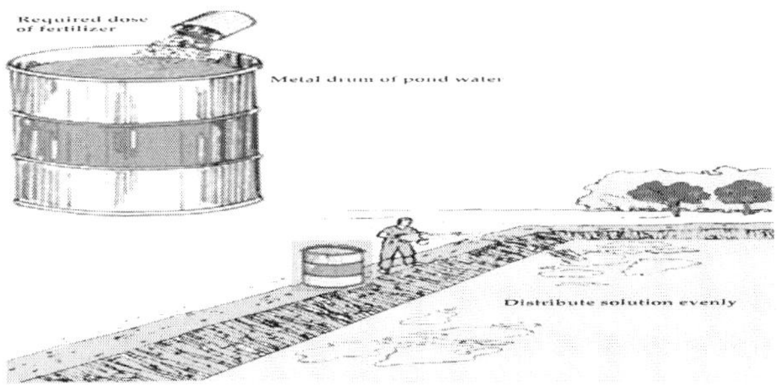


**Figure 5.4**: Organic fertilizers are normally used to help in the growth of organic foods in the fishponds.

*Source: FOA of the unit*

Semi-intensive fish farming is something in between intensive fish farming and extensive fish farming. In this case, the fish are known to obtain sufficient food from the natural web but in addition, they are also provided with supplementary food. With this, the fish get to grow faster and get to be ones of very large sizes given the fact that, they are provided with natural feeds and supplements. Apart from growing faster and larger, they tend to grow at a greater stocking density. The feeds of these fish vary as they may be of vegetable origin, fishmeal, other fish, or fish oil. Grass carp is one species that s normally farmed using this method and provides the highest global production tonnage. This fish is mainly farmed in China in both intensive and semi-intensive systems (Assefa & Abunna, 2018). When semi-intensive farming is used, the grass carp are normally stocked in ponds or in pens with other carp species. The species that are put in the ponds in conjunction with the grass carp may be either the major species or the secondary species. In this case, the terrestrial grasses and the aquatic weeds form the main feeds for the fish. In order to solve the labor costs, the farmers may decide to use commercial feeds, which include pellets and vegetable by-products. The supplements are normally used when the growth of water grasses and algae is reduced.

The intensive aquaculture system is one that is currently in use in most parts of the world given the fact that food production by natural means is becoming an actual problem and this method forms as a solution for the same.

## 5.2 WHY INTENSIVE AQUACULTURE

In this system, more fish is produced from a small piece of land or place. The amount of fish that is produced per unit area is quite high as a small tank can produce quite a large number of fish.

This system is fully controlled by the farmer. The farmer controls each and every aspect of this system. Due to the fact that the fish are fed on commercially produced feed, the farmer controls the quantity and quality of feed that the fish are going to receive. The oxygen levels are controlled and monitored by the farmer as to the larger extent he/she can control the quality of water that is supplied into the fishpond.

It creates a variety of employment opportunities. Due to the fact that this system is closely monitored and there are several systems or machines in place, there has to be manpower that takes care of these machines. Some of the people that are required to are, a technician or mechanic to ensure that all the machines are in proper working conditions, a manager of control all activities in the farm i.e. feeding the fish, laborers so that they can carry out various activities needed in the farm.

**Figure 5.5**: Fish tank used to raise fish. The feed is given to the fish and water used is recycled after being used.

*Source: India exporters*

There is minimized introduction of toxicants and pollutants. The water that is normally used in these systems is usually purified then reused. The water after leaving the fish tank moves into the purification system where the wastes are removed and oxygen is added. After going through the purification system, it is pumped back into the fish tanks to be used by the fish.

The drawbacks that are involved in aquaculture are minimized. These may include sharing of resources, elimination of certain species, diseases in fisheries and other natural habitats, aquaculture being a risky venture are some of the drawbacks that intensive aquaculture is trying to eliminate (Jana et al., 2018).

There is little or no harm to the environment. The water used is recycled and some intensive farms use the water to carry out hydroponics hence utilizing the nutrients that are found in the water. The water used is not dumped into other water sources as there is recycling and reuse of water.

## 5.3 DIFFICULTIES OF INTENSIVE AQUACULTURE

It is impossible to produce fish that are of big sizes in this system. This system allows for the keeping of a large number of fish in an enclosed tank. Due to this, the fish will always be competing for feed and this will mean the fish will not be able to grow in size. Due to the limited space and fish, being in small round tanks, the large fish cannot be stored in these fish tanks.

**Figure 5.6**: Tanks used, as fishponds can be disadvantageous especially if the fish are big.

The cost of input when compared to extensive farming per the weight of fish is quite high. The fish kept in this system are too many in each tank and the feed that is used is a lot. The quantity of feed used is high which means that the cost will increase. The fish that are harvested are high in number but the weight of the fish is quite less.

There is a lot of expenditure on labor. This system requires a lot of monitoring and it also requires a lot of skill. This means the individuals that are employed have to be skilled in various fields. There need to be technicians, mechanics, managers and aquaculture experts. There also needs to be unskilled labor to carry out the various tasks on the farm. With all these individuals, it means the budget for wages of the employees will be quite high.

It has a high cost of maintenance. There is a lot of machinery that is used in this system. With all the machinery that is used it means they have to be constantly maintained in order for them to be working properly and to minimize or prevent any losses from occurring. The machines that are used are quite expensive and skilled individuals have to maintain them and check them regularly, especially in large intensive aquaculture farms.

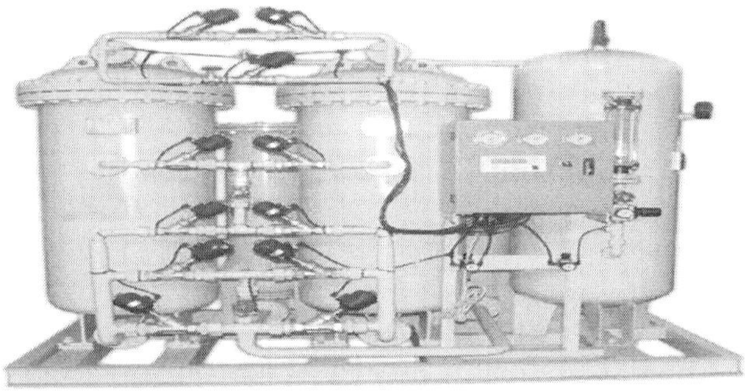

**Figure 5.7**: An oxygen generator. The machine is complicated and requires a skilled technician and constant monitoring.

It provides a non-natural environment for the growth of fish. The area or place that the fish grow in a constructed habitat. The fish at no point in their lives do not come in contact with any natural habitat. Many intensive farms produce their own fish, which means they have their own hatcheries (Hossain et al., 2020).

Water in this system has to be monitored on a regular basis and it also has to be recycled. The water that is used is monitored as some types of fish only reproduce in certain temperatures and also in certain pH. Due to all these factors, the water that is used has to be checked and measured regularly. The tanks are fitted with sensors to monitor temperature, pH and water clarity. The water that is used is purified and recycled or stored to be used later.

The water has to be treated mechanically so as to prevent deterioration. The water that is used contains a lot of waste and feed that has not been taken up by the fish. The water has to be treated to remove the wastes and the other minerals in the water to give the fish a good environment to live in. The volume of water and the fact that the water has to be of certain quality it has to be treated by machines to achieve the desired quality and save on time.

With an increase in fish being consumed in the world and the ever-growing demand for white meat, it simply means the demand for fish will increase rapidly. With all this, there will be more intensive farms coming up. The continent is looking for ways to produce fish without causing harm to the environment and with intensive farming carrying out recycling and reusing its water and purifying it. It means the environment has been protected.

## 5.4 COMMERCIAL AQUACULTURE PRODUCTION TECHNOLOGY

Murray cod is a highly sought endemic Murray –Darling finfish second to table fish. Generally, it has various ranges of uses such as for commercial or economic purposes, recreational as well as conservation uses. Recently, the Murray cod has supported enormously lucrative including the very small commercial fishery. The technological developments are newly discovered every day and they have helped increase the production as well as grow the industry concerned with Murray cod. From the 1970s to the early 1980s various breeding techniques for Murray cod have been established and improved and as a result ,thousands of fingerling Murray cod have been produced in fish hatcheries daily. The number of enterprises involving in Murray cod production has been on the rise due to the major technological developments that have taken place in the industry as they led to ease production of Murray cod (Balcázar & Ma, 2007). The fish have been used in the past for stocking public and private waterways for recreational and conservation activities worldwide. Today, there is considerable interest among producers and marketers to outgrow the production of Murray cod. This is so as to satisfy the growing market demanding the local and international markets. The trend has been as a result of the favorable information of increased technological development in the production activities as well as the associated increasing consumer demands.

Over the last 30 years, the techniques for especially large-scale production of Murray cod, also known as *Maccullochella peelii* fingerling have been the upscale. Traditionally, the fingerlings were produced for restocking public and private waters but currently, the uses have been quiet enhanced. The technological developments over the years have led to the increasing trend towards weaning and on growing Murray cod for food all over overseas including Australia. Various departments have been at the forefront in support of the technological advancement of the production processes for many years. Many research and developments as well as projects have been carried out in line with Murray cod production.

Genetics and reproduction-enhanced technologies have been developed, demonstrated, evaluated and implemented in the Victorian aquaculture sector in a bid to enhance the production of Murray codfish. This has led to increase in the numbers of fingerlings produced in the industry. The increase is mainly attributed to the enhanced production efficiencies .besides, value addition has been on the rise for the Murray cod fingerings. This has help increase as well as retain the market access and viability of the goods. The practice change in genetic and reproduction technologies of Murray cod and investments in genetic biotechnologies has been evident. This has been so through the development and implementation of strategies of the targeted route to market or end-user and for commercialization purposes.

**Figure 5.8**: Murray cod industry development plan aquaculture water resources.

*Source: Scribd*

A revolution in the application of genomics and gene-related biotechnologies in aquaculture and in the agricultural sector as a whole has been evident in the recent past. Biotechnology research and development have been used to increase production efficiencies of the various activities and outputs produced in agriculture .they have led to reduced production costs. The manipulations of the genes and chromosomes of cultivated species have been on the rise for most agricultural crops and animals. In wild fisheries, biotechnology has been applied to improve the management in such as the DNA being used to separate or differentiate the various populations and in the management of captive breeding programs for stock enhancement or replenishment. A major biotech project has been funded by the state government of Victoria in Australia. The technology is an approach that has led to the increase in the aquaculture and enhancement of fisheries management in the industry, especially of the Murray codfish. There has been a challenge , however, of applying advanced genetic and reproduction technologies to the high-value inland species so as to increase the production efficiencies as well as manage wild populations.

The experienced production performance gains coupled with the on going reliability in Murray cod production for the next five to ten years or so could see the expansion of the Murray cod production industry as well as the aquaculture farming in Australia other regions internationally. The increased applications for the various technological improvements in the enterprises would provide extra benefits for the management and control of activities of the native fin fish biodiversity (Balcázar & Ma, 2007). This is especially for the Murray cod species variety which has been on the decline in the wild while is continuous to be kept for several multiple purposes such as commercial, recreation as well as conservation. The progress made in Murray cod in Australia, especially has been quite enormous. This is the largest indigenous freshwater fish in the region. It is an iconic species with various associated values among them including cultural values. However, in the area overfishing, habitat loss and modification within the Murray cod natural range has been the order of the day. This has seen the reduced numbers of cod since the early 1900s.

The Murray darling basins in Australia mainly entail a network of streams, rivers and water bodies that are within the catchment of over one million square kilometer catchment areas, which is approximately 14% of the land surface of the main land. The fish variety has been listed to be very vulnerable under the Australian legislation environment protection and biodiversity conservation act 1999. This act is designed to protect the conservation status of Murray codfish. Over about 11 million hatcheries, reared juveniles have been released in the wild since the 1980s. This has been so as to enhance the recreational species of fisheries and in addition to the conservation purposes. The Murray cod is widely known for its excellent aqua cultural prospects and the availability of markets as small quantities are already sold into the local as well as the domestic markets worldwide. The associated attributes of the Murray cod include the establishment of hatchery production of fry, the acceptability of readily artificial food the physical handling hardness and the combined fast-growing rates as well as the ability of the fish to tolerate high stocking densities .in addition the high demand, especially among the Asian communities has led to the Murray cod enhanced features.

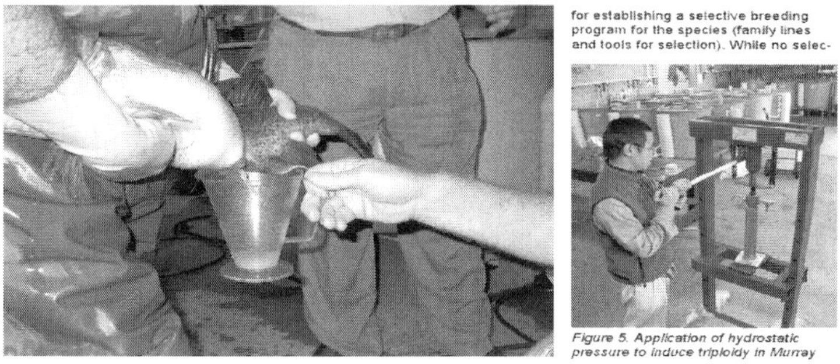


**Figure 5.9**: Hand stripping of Murray cod eggs.

The Murray cod requires a dietary protein nutrient of about 50% for enhanced and optimal growth of its performance. The species has the ability to effectively digest the food industry by-products such as defatted soybean meal, shark meat meal as well as meat meal. However, the protein generation abilities of the Murray cod have been limited to about 32 % of the fishmeal content in Murray cod diets. This could be replaced without compromising the growth performance of the fish. Information has been provided on cost-effective diets for intensive Murray cod culture for Murray cod culture. The technologies have been successfully tested under commercial conditions and are seen to be relevant in increasing the production of Murray cod. The new information involves adoption of a diet. The adoption of the diet for growing out of Murray cod in the systems may significantly reduce the associated costs in the overall production of the Murray cod fingerlings (Hemaiswarya et al., 2011).

The values for 16 water quality parameters were summarized to have the ability to provide a guide to provide the acceptable ranges to be used in intensive Murray cod aquaculture production activities. The contents of the simplified nutrient mass balance include a meal that was developed to estimate the amounts of the required nutrients, which are nitrogen and phosphorous discharge from a technologically developed Murray cod aquaculture system. For every tone of market size Murray cod that is harvested approximately about 12.4 kg phosphorous per ton and about 67 kg nitrogen per ton are to be based on the median nutrient input contents. The various changes in the FCR have brought significant feed influences in the nitrogen and phosphorous feed contents amount applied. Further, researchers have engaged themselves in the studies involving future construction, operations and management of the RAS systems for the production activities in the Murray cod in the fish industry.

Technological developments in commercial Murray cod production actives have also been brought in to deal with fish diseases and fish health issues affecting Murray cod. There are various strategies that have been developed so as to assist the various cultivators in the management of the health and well-being of stock in intensive aquaculture facilities. The various technological developments have been made on the water supply sources and treatments, quarantine and management of the new stock, management of hygiene in aquaculture facilities as well as in the monitoring of stock by fish and the identification of the health problems and use of drugs and chemicals in the production of Murray cod enterprises. The decision that has been deliberated on support pathway was presented so as to assist farmers in the characterization and the ease identification and management of major health problems in Murray cod aquaculture activities.

Murray cod products are today routinely cultured as are of various varieties for the use in the range of markets ranging from post-larvae and the fingerlings for stock enhancement and to promote it to the table fish that is sold at 0.6 to 4.0 kg .It can be live, fresh, grilled or cooked. Additionally, they can be sold as fillets into a number of markets. The fingerlings are mostly exported for growing reasons as Murray cod products have been sold in the international markets. Technological advancements in the sector have also led to improved market chain efficiency through more reliable and routine supply products of known qualities. Although the prices of Murray cod have been on the decline from about $24.00/kg in late 1999 to around $13.50/kg, information development on the marketing of Murray cod as from an industry standpoint has been enhancing. Cooperatives have also been on the trend. Production and economic efficiencies were investigated for five Murray cod RAS (1, 5, 25, 50 or 100 ton/annum). The methods used allowed that production data from the industry were combined with the results from the recent studies and were later analyzed using the aqua farmer feasibility software in order to determine and as well as evaluate the bioeconomic performance of each system (Hasimuna et al., 2020). The report indicated that on-farm diversification of the activity would improve the viability of small farms involving in Murray cod as there were significant opportunities for improved risk management strategies to be used in the larger systems for which the economies of scale could be achieved.

**Figure 5.10**: Murray cod can be very large and is a very expensive fresh water fish.

*Source: Fly lords Mag*

Murray cod scientifically known as *Maccullo chellapeelii* being a highly-priced fresh water fish has various advantages over the Chinese mandarin fish. It has gained popularity as a suitable species among the fish for its intensive culture. It is compared to the Chinese mandarin fish currently and even superseding. The Chinese mandarin fish was in the recent past considered to be the most or highly valued warm water fresh water species to be existing worldwide that was produced and cultured for commercial purposes. The associated advantages to the Murray cod farming are brought bought by the ease of the weaning to dry feeds, unlike the mandarin fish which only feeds on live fish particularly. The commercial culture of Murray cod is on the rise recently although the processes of cultivating have been artificially propagated in the recent past. This has mainly been due to the stock enhancement activities. The advancement of the production processes in Murray cod production has been increased and the biology documented.

Mostly, in most places, Murray cod are farmed in intensive, indoors systems that are recirculation as well as in outdoor ponds. The commercial diets formulated for Murray cod are however not available due to the species being new in the market. Majorly the farming has not been for commercial purposes as for the other species, which have dominated the market such as the salmonids that are particularly Atlantic salmon *Salmo salar* as well as barramundi species *Lates calarifer* that are available for dietary purposes to humans. In recent research and development projects, there has been one nutritional related study of a form, which showed that the Blue sac syndrome that could be observed in artificially propagated Murray cod larvae would have been as a result of deficiencies of essential and imperative amino acids required for the growth of Murray cod. For a long time, there has been commonly observed old brood stock that affects the Murray cod production. Nutritional requirements of the fish species differ widely for the various species, with carnivorous species particularly having relatively higher dietary protein requirements as compared to the omnivorous type of fish. The fish species differ in their lipids dietary requirements contents as well as the ability to make use of the carbohydrates as an energy source. Feeding costs for fish production account for about 30 to 60 % of the costs that are recurring of culturing carnivorous fish species. The fish are basically produced under intensive conditions in which nutritional requirements are essential (Freitas et al., 2020).

Murray cod is among the fish variety with the ability to tolerate relatively high levels of SBM in their diets. However, this has not been experienced in the BM. A study conducted by Fowler and Banks in 1976 found out that when the FM was replaced by the spray-dried BM there were no negative effects that were observed red in the inch of nook salmon up to 5% inclusions in the BM. The pathological effects were observed however, when the FM was put up to 17.55 as the spray-dried BM in the diets. The responses of different fish species to BM incorporated diets have been variable, the majority of the species were seen not to be able to tolerate the levels exceeding 20% in the diets.

**Figure 5.11**: Hybrid Murray cod is a technological development in the aquaculture industry.

*Source: Fishing world*

Another technological development in Murray cod has come about by how the hybrid fish have been used in recent years in aquaculture activities in a bid to increase the hybrid vigor. The processes carried out have been used to transfer or combine desirable traits between species in the efforts of reducing unwanted reproduction through the production of sterile offspring activities. Hybridization has been successful when induced in both the direct and reciprocal crosses among Murray cod and trout cod also known as *Maccullochellam acquariensis* scientifically. The hybrid between the two is however rare but they have been made possible through the hybrid systems. They at times occur naturally in the wild areas where both species are found sympatrically. Increasingly small numbers of hybrids are being farmed in the ongoing innovation practices in order of evaluating the aquaculture performances for example in evaluating the growth and environmental tolerance of the fish. The procedures have also been essential in evaluating to determine the Murray cod fish fertility status.

The Monash Institute of Medical Research (Melbourne) in collaboration with other private entities has been investigating while carrying out research on strategies for cryopreserving sperm in liquid nitrogen as a way of storing genetic materials. This is because the strategy need should be cost-effective for storing as well as to be used in the conservation and selection purposes of various breeding programs. Evaluations have been carried out and trials have been identified on cryoprotectant composition and freezing methods to the Murray cod sperm species. This has so been in the labors of identifying the Murray cod sperms that could provide more levels of sperm motility post-thawing. During the fertilization processes, the trials have managed to achieve hatch rates of up to 65% of the hatch rates contain eggs that have been fertilized by fresh viable sperms (Faruk et al., 2004). Technologies that would lead to early detection of sex have been important in evaluating the reproduction technologies that would in place affect or manipulate sex so as to increase the productivity of the Murray cod .nevertheless, breeding of the Murray cod cannot be accurately sexed outside the breeding season without first applying some intensive techniques whilst juveniles cannot reproduce at all. In addition, there other technological advancements in the fish sector of the Murray cod production processes such as the application of hydrostatic pressures. The hydrostatic pressure in Murray cod production is used to induce triploid in the eggs for multiple productivity purposes. The technology was developed and has been highly adopted in the world for the increased production of aquaculture.

Options to improving the profitability of the production of Murray cod have been made to include the adoption of more energy-facilitated systems as well as labor-efficient RAS technologies. Improved husbandry to optimize the growth of the Murray cod as well as its survival has also been kept into considerations. This is as farmers are being encouraged to use more cost-effective diets as well as use the genetically improved seed stock from the developed technologies so as to enhance their productivity at the farm levels.

There have arisen concerns that should be based on future research and development purposes to the industry such as the improvement in the production systems of Murray cod as there exist several challenges. The system should be developed to ease the production processes and should be in association with the IAAS. This is in the matters of advancing the genetically modified varieties of Murray cod fish available. Guidelines should be developed for procurements of the wild fish and in the management of brood fish as well as the various breeding programs for the production and growth of seed stock that is used for stock enhancement activities. the genetic and biotechnologies developed should have the ability to establish selective breeding programs so as to ensure long-term genetic integrity as well as enhancement of the cultured Murray cod stocks. Additionally, technological advancements should be made on environmentally cleaner diets that have limited or reduced phosphorous and nitrogen contents.

## 5.5 DEVELOPMENT OF CONTROLLED INTENSIVE AQUACULTURE WITH LIMITED WATER EXCHANGE AND ADJUSTED CARBON TO NITROGEN RATIO

**Figure 5.12**: Aeration of fish ponds is important so as to ensure better growth of the fish in the ponds.

Intensive fish aquaculture can only be made possible when there is proper aeration and do mixing in order to avoid the creation of anaerobic sites. When one does water exchange, one has the capacity to flush out some of the toxic components that may affect the fish. During aeration, water is normally drained which helps in relieving growth retardation. In this case, the ponds are fully drained then the ponds are fully aerated for a couple of days before the ponds are restocked again. This shows that growth retardation of the fish that are normally kept by use of intensive care is normally brought about by the reduction of oxygen due to the accumulation of sediments. As such, the need for aeration and water exchange will help in the growth of the fish. This has been one development in intensive aquaculture that has proven very beneficial. Since intensive aquaculture requires that the fish be commercially fed, the growth of seaweeds and planktons in the pond may be overwhelming for the fish. The plants in the ponds may be too much that it sucks up some of the oxygen that may be needed by the fish. As such, with this method, the farmers can be able to remove the plants and some of the sediments at the bottom of the pond. As such, it is important for the farmer for this type of system needs to have more than one pond to enable him to perform a transfer to the other pond so as to allow for water removal for few days to allow for aeration (Deb et al., 2021). This is one thing that makes this method expensive given the fact that that one needs to always ensure that the ponds are aerated. The process should be done in a very careful manner given the fact that the changes in scenery may prove to be very stressful to the fish, which may lead to losses. Proper aeration of the pond is important as it ensures that the growth of the fish is going on well and it does go on faster given the fact that it is an improved way of marketing fish.

## 5.6 DEVELOPMENT OF SLUDGE REMOVER FOR INTENSIVE AQUACULTURE

In pond fish production especially in the ponds, there are a number of technical constraints that are normally brought about by abiotic components that include the problems with soil, temperature, water among others and biotic components that include diseases, predators and pests. These biotic factors and abiotic factors are known to be the major causes of financial loss in the areas that practice intensive systems of aquaculture. One major abiotic problem is the aquatic sludge that deposits at the bottom of the pond. In fish farms, sludge are known to originate from three things including drum filters, bio filters and fish feces. Management of the sludge and dewatering of the ponds have however proven to be great challenges to fish farmers. For the desirable growth of fish to take, place: there needs to be the removal of sludge from the ponds. Ideally, the formation of sludge in the ponds does not only affect the fish in the water ponds but acts as a pollutant to the environment. The sludge that is formed from the ponds in most cases is partially composed of some of the uneaten food, bacterial growth in the pond and stabilized excreta.

**Figure 5.13**: Sludge needs to be removed from the fishponds to ensure proper development of the fish in the ponds.

*Source: Pond Aquarium Problem Solver*

Reports from researchers show that, regardless of the type of fish that is being bred in the fishponds, the breeding of fish in the presence of sludge will be the downfall of the fish production system. As a matter of fact, human beings that will consume the same fish may have contamination. Ideally, it is neither appropriate nor healthy for an individual to consume contaminated fish. Furthermore, the accumulation of solids within the aquaculture tanks and systems, it may prove to be suitable conditions for the growth of fish pathogens. As a matter of fact, when these solids are not removed as soon as possible, they can break down into smaller particles. When this happens, the nutrients in the water will be destroyed, the water quality in the pond will be degraded and there will be high demand for oxygen in the ponds (Dauda, 2020). With the formation of bacteria in the ponds, the fish will have limited access to oxygen in the stream and hence they will be subjected to stress that may retard their rates of growth. With the stress in oxygen demands, there will be increased dissolved carbon dioxide levels. The small-suspended solids tend to cause gill irritation. As such, there will be a reduction of immune system efficiency, which then causes disease outbreaks in the pond and as such losses for the farmer. As such, it is important for the fish bearers to always remove the solid wastes as rapidly as possible so that they can be able to positively affect the health of the fish species in the ponds. Researchers claim that it is important to come up with a sludge removal system to ensure easy and efficient removal of sludge from the fishpond.

There are some farmers that are using the chemical, physical and biological methods that are normally used in wastewater treatment to treat their ponds. These methods are known to be very suitable for tanks and small ponds. When it comes to large and commercial intensive aquaculture, these methods can prove to be inefficient and inadequate. As such, the mechanical method may prove to be more adequate for the removal of sludge. For the case of farmers in Vietnam, a sludge pump is one that is used for the removal of sludge from the fishpond. They use these pumps for the purpose of trying to make the fishpond environmentally friendly and healthy for the fish. These pumps are generally used at the bottom of the fishponds to suck out the waste and the sunken debris from the bottom of the ponds. They are normally attached to a discharge tube and a power source. Sludge pumps are not that common in most parts of the world but it has proven to be very helpful especially after the establishment by the fish farmers in Vietnam.

It is very important to understand how the sludge remover works. The additives of the sludge remover are carried at the dyke of the pond where sludge remover is to be operated. At first, drift is positioned within the pond water. Then pump, engine, crane, propeller, rudder, and transport pipe are geared up with the drift. Two operators are required to operate the sludge remover. One man or woman operates the machine, driving at the drift and any other man or woman manages the sludge pipe from the dyke of the pond. Sludge pump with sucker is adjusted in line with the intensity of water. The engine is started and the combination of sludge and water comes out via transport pipe to the sludge deposition pit or tank. The amount of sludge may be elevated or reduced via way of means of decreasing or lifting the sludge sucker via way of means of the lever of the crane. When the location of the sludge sucker is adjusted then propeller is started via way of means of the lever (Dauda et al., 2019). Then the drift with sludge pump actions forward. The velocity of drift can be elevated or reduced via way of means of adjusting the propeller lever. The route of the motion is managed via way of means of the rudder. Sometimes it is far tough for turning the sludge remover the use of the current rudder. Then the drift is made to become round with the assist of a bamboo pole (normally utilized in a boat). The position of the transport pipe is adjusted via way of means of any other operator. In this method, sludge from the lowest of the pond starts to be eliminated from one facet of the pond and be completed as much as any other end. When the sludge deposition pit or tank is stuffed up any other pit or tank is used. This technique is sustained until the elimination of sludge from the complete pond is completed.

This shows that the operation of the sludge remover is quite easy for individuals to work through. All these developments in the intensive aquaculture have proven that they can be able to help the farmers in their production of fish, especially for commercial purposes.

Chapter 6

# FEED MANAGEMENT IN INTENSIVE AQUACULTURE

## CONTENTS

6.1 Synoptic Review of Farm-Made and Commercial Feed Use ............ 137

6.2 Feed Production and Handling: Feed Formulation .......................... 140

6.3 Manufacturing Technologies ........................................................ 143

6.4 Feed Transport, Storage and Handling ............................................ 144

6.5 Top Dressing and Feed Additives .................................................. 145

6.6 Optimizing Feed Management Strategies ........................................ 146

6.7 The Role of Natural Productivity and the Implications For Feed Management ................................................................ 150

6.8 The Use And Efficacy of Feeding Devices ........................................ 151

6.9 Feed Management and the Environment ......................................... 153

6.10 The Economics of Feed Management ............................................. 154

6.11 Aquafeed Policy, Regulations and Governance ............................. 157

6.12 Training and Developing Information Networks to Improve Feed Management ......................................................... 159

Feed expenses generally account for around 40 and 60 percent of production costs in semi-intensive and intensive aquaculture systems. It is vital that farmers have access to good quality feed at affordable prices to assure profitability and improve the utilization of their feed by establishing suitable on-farm feed management procedures. The paper provides an introduction of feed production and management strategies in farming, and provides an outline of the key difficulties and limitations in improving feed output. The analysis is based on results from country case studies specific to species, regional reviews and the current expert topic reviews.

Extensive fish farming refers to fish farming in medium- to large-sized ponds or water bodies where the fish production is solely dependent on the water's natural productivity, which is only marginally or moderately improved. Inputs from outside sources are limited; costs are kept low; capital investment is minimal; and the amount of fish produced per unit area is little. In a nutshell, the level of control over the production factors is kept to a minimum. The rate of return on labor is very high.

**Figure 6.1**: Feed distribution using an automatic feeder for cage culture of Atlantic salmon near Bodo, Norway.

*Source:* http://www.fao.org/3/al363e/al363e21.pdf

Intensive fish farming, on the other hand, means that a large number of fish are produced per unit of raising area. Production parameters including feed, water quality, and the quality of stocked fingerlings are all regulated to improve the culture's production circumstances. Throughout the production cycle, there is constant monitoring. All of these controls, of course, necessitate high-tech procedures and capital-intensive investments, which raise manufacturing costs. Increasing production costs must be justified by increased profits. Natural productivity makes a minor or non-existent contribution to fish output. Aside from that, intensive fish farming is associated with high expenses and environmental risks (Bossier &Ekasari, 2017).

Aside from these two types of fish farming, some people talk about semi-intensive fish farming, which refers to techniques that include features of both. This, on the other hand, is ill-defined. Before recommending one of these kinds, it's important to consider the circumstances in which fish farming is carried out. What are the goals of the farmer? The primary goal is to ensure his family's food security and income. This can be achieved in two ways: first, by cultivating food crops, and second, by diversifying the farmer's operations across a variety of agriculture and non-agriculture endeavors in order to achieve his income-generating goals.

The rationale is to manage risk (pre-emptively) and deal with loss (ex post). The assets (resources, human capital, know-how) and investment capacities will have to be considered in the plans devised. Optimal market circumstances must be present in order for him to carry out his strategies and achieve his goals. Essentially, these are market systems for agricultural and non-agricultural factors and products. Food or labor shortages, for example, can affect market performance and prohibit a farmer from attaining his goals. In order to support his methods, the farmer seeks to maximize the utilization of resources within his reach, putting unused or underutilized resources to good use.

As a result, the expected return on investment in fish farming is the objective reason for getting into the business. The decision to engage in income-generating activities is based on the projected returns and risks associated with each alternative.

## 6.1 SYNOPTIC REVIEW OF FARM-MADE AND COMMERCIAL FEED USE

Feed selection and management strategies have substantial effects on the financial performance of a production system because feed is one of the biggest operational costs in aquaculture systems. Farmers' feed input choices are influenced by a variety of factors, including the market (local, export), the value of the fish, the farmer's economic ability, the species, the culture system, and the volume of production. Fertilizers, farm-made feeds, or locally manufactured small-scale commercial feeds including one or more ingredient supplies are frequently the only inputs for low-value species cultivated for local consumption.

Indian big carp farmed under extensive or semi–intensive conditions in India and Bangladesh are an example of these systems. Commercially made pelleted feed inputs, on the other hand, are employed for high-value species cultivated in intensive systems. Salmonids in Europe and the Americas, as well as high-value marine finfish (e.g. groupers, barramundi, and snappers) that are increasingly being produced in Southeast Asia are examples. Intensification pressures are increasing as competition for land and water resources becomes more intense.

The use of nutritionally balanced feeds is required as farming methods become more intensive, which is increasing demand for both farm-made and commercially manufactured feeds. This progressive movement toward intensity was evident in the case studies. In Egypt, for example, greater demand for industrially processed feeds has resulted from the intensification of pond culture practices for Nile tilapia (*Oreochromis niloticus*) and the incorporation of intensive cage culture technology, which now account for 60% and 100% of the feed used in these segments, respectively. As a result of the increased demand, the country's commercial feed manufacturing capacity has increased from 20 000 tonnes per year in 1999 to 420 000 tonnes per year in 2010 (Bostock et al., 2010). Similarly, increased demand for commercially manufactured feeds has come from increased production of black tiger shrimp (*Penaeus monodon*) in India, Nile tilapia (*Oreochromis niloticus*) in the Philippines, and striped catfish *(Pangasianodon hypophthalmus*) in Vietnam.

Commercially manufactured feeds were reported to be accessible in eight countries where research findings were completed (Bangladesh, China, Egypt, Ghana, India, the Philippines, Thailand, and Vietnam), and all but Ghana reported commercial manufacturing capability. According to a study, locally produced commercially produced feeds are only available in five Sub-Saharan African countries, indicating that, unlike other producer regions (Europe, Asia, and the Americas), Sub-Saharan Africa has limited access to locally manufactured feeds and must rely on farm-made feeds or imported industrially processed feeds.

**Figure 6.2**: Feeding striped catfish (Pangasianodonhypophthalmus) with farm-made aquafeed, Mekong Delta, Vietnam.

*Source: http://www.fao.org/3/al363e/al363e21.pdf*

While industrially processed feeds are frequently tailored to meet the nutritional needs of specific species and life stages, farm-made feeds are typically made out of simple materials that are supplied as simple mixes, doughs, or compressed pellets, dependent on the culture system. While the quality of farm-made feeds is determined by the formulation used, the quality and availability of components, and the production procedures, they are often less expensive than commercially manufactured feeds and continue to be the principal feed supply for many semi-intensive farms.

The cheaper price of farm-made feeds allows resource-strapped farmers to purchase small amounts of feed at a time, boosting affordability and allowing them to efficiently assess their on-farm cash flows. Farm-made feeds are becoming increasingly important to maintaining sectorial growth as farming systems become more intensive and demand for good quality, cost-effective farm-made feeds grows. It has been noted that one way of improving aquaculture production is to improve the quality of farm-made feeds. The quality of farm-made feeds has already improved significantly in some agricultural sectors.

Farmers in Vietnam, for example, who utilize farm-made feeds to raise striped catfish have enhanced their feed compositions and manufacturing procedures. The feeds are processed to generate semi-moist pellets with better water stability, and the formulas now incorporate up to six ingredient sources. Farmers' access to data (e.g., formulas, ingredient availability, and costs) has improved in Bangladesh as a result of the growth of farmer associations, as well as their purchasing power and access to production technologies (Corsin et al., 2008). Farmers now have access to small-scale commercial feed producers and suppliers who are able to develop feed batches in relatively small amounts, allowing them to negotiate a better deal for ingredient prices and buy in bulk; providing a level of manufacturing technology and feed quality that would otherwise be unattainable. In this context, the possibility for promoting public-private partnerships between small-scale feed manufacturers and farmers/farmer associations as a vehicle to expand access to cost-effective, high-quality feeds should be looked into further.

## 6.2 FEED PRODUCTION AND HANDLING: FEED FORMULATION

**Figure 6.3**: Preparation of farm-made supplemental feed using locally available feed ingredients, Mymensingh, Bangladesh.

*Source: http://www.fao.org/3/al363e/al363e21.pdf*

The provision of nutritionally balanced feeds to farmers is a precondition for cost-effective production. For both commercial and farm-made feed producers, formulation concerns, particularly the supply of species-specific feeds that fulfill the nutrient needs of the several life stages of the farmed animal (larval, fry, grower, finisher, and broodstock), remain a problem. Many of the industrially processed formulations accessible to farmers are based on laboratory formulations with high-quality components; however, only a few are tested under commercial farming situations. Formulations developed purely in the lab do not necessarily transition well to commercial environments, where lower-quality feed ingredients and low-cost formulae are used. Similarly, the formulas are not often backed up by extensive scientific study, are poorly formulated, and are supplied to farmers who may be unaware of their farmed species' nutritional requirements. Experts have pointed to this issue in Southeast Asia's growing marine finfish industry, where farmers were encouraged to use the same commercial formulation for barramundi (*Lates calcarifer*) and grouper cultivation, despite the fact that the two species have differing nutritional needs.

In fact, improper formulations were discovered to be a widespread issue across a variety of industries. Tilapia farmers in China, for example, have been reported to utilize commercial grow-out formulas with higher levels of dietary protein than needed, while tilapia farmers in Thailand have been documented to use commercial grow-out feeds developed for catfish and carps on a regular basis. Striped catfish producers in Vietnam reported utilizing grow-out feeds with protein levels that were less than ideal for the higher size classes. Despite the availability of whiteleg shrimp (*Penaeus vannamei*)-specific meals, farmers reported utilizing feeds developed for black tiger shrimp (*Penaeus monodon*), which have a larger protein content than the species requires. Feed inefficiencies and higher production costs will inevitably follow from the use of poorly formulated feeds that fail to meet the nutrient needs of the species and their various life phases. Farmers, feed suppliers, and unregulated feed producers all need to be informed about the necessity of selecting and supplying proper species and size-specific formulas (Bondad-Reantaso et al., 2005).

Some commercial formulas for specific life stages (for example, fry, fingerling, and grower) are so identical in terms of nutrient content that they are nearly indistinguishable. This raises a number of questions, including the nutritional rationale for the formulas, the correlation between nutrient inclusion levels and cost differentials, and the manufacturers' capacity to maintain such modest variances in nutrient levels in the feed on a consistent basis. Resolving these challenges and ensuring that farmers receive proper formulations at cost-effective pricing should be considered a top priority, as it would undoubtedly improve the farming operation's economic performance.

**Figure 6.4**: Fish nutrition, feeds, and feeding management play important roles in increasing the productivity of aquaculture farms.

*Source:* https://repository.seafdec.org.ph/bitstream/handle/10862/2158/MillamenaOM2002-Introduction-to-nutrition-in-tropical-aquaculture.pdf?sequence=1&isAllowed=y

While substantial study has been done to determine the nutritional requirements of different species groupings, much of this information has not been disseminated to farmers who make farm-made feeds or small-scale feed manufacturers. Many farmers that produce farm-made feeds are apparently uninformed of their farmed species' nutrient requirements, particularly dietary protein and energy levels, and how they alter over time. Formulations are frequently based on previous experiences (what the farmers have discovered to work), availability and pricing of feed ingredients, and recommendations from other farmers and feed component suppliers.

Farmers of tilapia in Thailand, for example, said they lacked basic information that might be used as formulation templates. Simple training manuals focusing on the nutritional requirements of farmed species, the availability, quality, composition, and cost of local feed ingredient sources, and methods to formulate feeds that meet specific nutritional requirements would greatly improve the quality of their formulations, as well as the economic efficiency of their farming operations. Later on, we will go over the training and information dissemination needs in greater depth.

## 6.3 MANUFACTURING TECHNOLOGIES

In Asia and Africa, most aquafeeds are manufactured on-farm or by small-scale semi-commercial feed businesses. Increased productivity and cost reductions are projected to result from improvements in the quality and preparation of on-farm feeds. The production procedures and kind of feed produced can have a substantial impact on feed performance, regardless of the quality of the feed ingredients used or the formulations utilized. While most farmers understand the need of using high-quality feed ingredients, they don't always realize that feed preparation has a substantial impact on feed quality and usage.

**Figure 6.5**: Production of high quality feeds starts with good quality ingredients and proper quality control throughout processing until use in fish farms.

*Source: https://repository.seafdec.org.ph/bitstream/handle/10862/2158/MillamenaOM2002-Introduction-to-nutrition-in-tropical-aquaculture.pdf?sequence=1&isAllowed=y*

Many of the feed ingredients used in farm-made tilapia feeds in China, for instance, are inadequately milled and do not meet the feed process standards highlighted in the national feed guidelines. Feeds presented as simple dry or moist mixtures or as moist mixed feeds are disseminated in the water column, resulting in low intake rates and high economic feed conversion ratios (eFCR). Farmers can improve feed efficiency by using basic extruders and compressing their feed materials into dry pellets. Similarly, enhancing pellet milling and binding qualities minimizes fines, increases pellet hardness and water stability, improves eFCR, and saves money for the farmer (Barman et al., 2013). Focusing on farm-made and small-scale feed manufacturing efficiencies is anticipated to result in significant advances in on-farm feed efficiencies. These industries are currently hampered by a number of problems, including a lack of access to capital, technological advancements, understanding of feed formulation and production, and training. It's worth thinking about the possibility of forming public-private partnerships with farmer organizations and institutions to pool resources and boost access to high-quality manufacturing capacity.

## 6.4 FEED TRANSPORT, STORAGE AND HANDLING

Farmers were largely unaware of the need of using suitable feed transport, handling, and storage practices, according to the majority of case studies. Since sea freight storage conditions are often inadequate and, depending on the route, delivery periods might be significant, imported industrial diets are especially susceptible to deterioration during transportation. Similarly, transporting feeds in open trucks, motorcycles, and bicycles can result in extended transit times and, on bad roads, damage to the pellets, as well as an increase in costs.

The majority of tilapia farmers in the Philippines were found to be inexperienced with feed storage concerns, and they learned about proper storage techniques only after encountering feed storage problems. Poor feed storage practices include everything from Ghanaian tilapia farmers storing their feeds in the open or under tarpaulins at night, subjecting them to moisture, pests, and weather conditions, to Indian shrimp farmers storing their feeds in metal containers and subjecting them to the sun and high temperatures. Nutrient loss, feed spoilage, poorer production, and poor economic returns can all come from improper feed storage conditions.

**Figure 6.6**: A nutritionally-balanced feed and adequate feeding are important factors that maximize fish production and profitability.

*Source: https://repository.seafdec.org.ph/bitstream/handle/10862/2158/MillamenaOM2002-Introduction-to-nutrition-in-tropical-aquaculture.pdf?sequence=1&isAllowed=y*

Long-term storage conditions that include light, heat, humidity, air, and water, as well as microbial/ pest infestation (bacteria, fungi, insects, and rodents) have a detrimental effect on feed quality. Feeds should be kept in cool, well-ventilated areas away from the weather, extremes of heat and humidity, and pests; they should also be used in a first-in, first-out manner. Modest-scale feed manufacturers and fish farms, according to experts, typically have small facilities that prevent them from purchasing raw materials in bulk when prices are low, increasing feed costs. Better management recommendations focused on feed storage and handling difficulties should be created and conveyed to farmers where appropriate.

## 6.5 TOP DRESSING AND FEED ADDITIVES

Top dressing, or the application of chemicals to feeds by farmers, usually necessitates spraying the final feeds with compounds that are incorporated into the feeds. This method is commonly used to add therapeutants, probiotics, or nutrients that farmers believe are inadequate or absent in their feeds, as well as feed attractants to increase intake. This technique has been recorded in a variety of industries. Probiotics and feed additives, for example, were reported by 75% of Indian shrimp producers, adding US$39/ tonne to the feed cost.

The source and grade of the chemicals used, as well as their cost and efficacy, are all key factors when it comes to top dressing. The management and use of therapeutants such as antibiotics and hormones are of particular significance. Aside from the legality of employing these compounds, if they are used improperly, they may cause animals to be dosed at sub-optimal levels, reducing efficacy, perhaps increasing disease resistance, and severely damaging the environment. Furthermore, if farmers get their medicines from shady dealers, they may end up utilizing illicit, adulterated, or low-quality drugs with low levels of active ingredients, or chemicals that are not fit for purpose (Benetti et al., 2021). Other issues include worker exposure to active pesticides and consumer safety as a result of farmers failing to follow the proper withdrawal durations. While many nations have adopted legislation to regulate the use of veterinarian medicines in aquafeeds, compliance is sometimes weak, and regulatory bodies are frequently ill-equipped to oversee their distribution and use. Another concern is the expense and effectiveness of the pesticides being used. Many farmers use additives in the mistaken notion that they are enhancing their production characteristics, however, only a few of these items have been scientifically validated for efficacy and cost-efficiency. It's unclear whether the farmers are merely increasing their production expenses or if there are real, tangible benefits to using them. Depending on the industry and country, solving these problems will likely necessitate a variety of interventions, such as farmer education, the development of improved management practices to advise farmers when and how to top dress their feed, as well as the legal status of the therapeutants and chemicals, regulatory controls and compliance, and research to determine the efficacy and safety of the therapeutants and chemicals.

## 6.6 OPTIMIZING FEED MANAGEMENT STRATEGIES

The profitability of a commercial farming operation is critical to the farmer's livelihood. Adopting effective feed management strategies is critical to ensure that feed utilization is optimized and that the farmer receives the maximum economic returns. While feeding to satiation will achieve maximum growth rates, over- or under-feeding will result in feed inefficiency and, in the event of over-feeding, increased farm waste levels. As hierarchies emerge, underfeeding expresses itself in reduced growth rates and increased size diversity in the population. Farmers must determine optimal ration sizes and feeding rates, feeding frequencies, and feeding periods that take into account the endogenous feeding patterns of the farmed species when optimizing feeding techniques. Farmers who use commercially prepared feeds are frequently, but not always, given feeding tables and technical assistance in setting ration sizes and feeding schedules, according to the case studies included in this document.

**Figure 6.7**: A good quality feed and appropriate feeding management maximize feed efficiency, lower cost of production, and reduce environmental degradation.

*Source: https://repository.seafdec.org.ph/bitstream/handle/10862/2158/MillamenaOM2002-Introduction-to-nutrition-in-tropical-aquaculture.pdf?sequence=1&isAllowed=y*

In many ways, it is in the feed manufacturer's best interests to guarantee that their feeds are used properly; it promotes positive production results for farmers and allows them to form long-term commercial ties. Farmers that believe they are receiving inadequate growth responses from a feed will switch suppliers rapidly. Farmers who use farm-made feeds and buy feed ingredients from suppliers are far less likely to have access to the data they need to figure out how to arrange their feeding schedules. Farmers will struggle to calculate suitable feed rations in the absence of this knowledge, and they are more likely to choose ineffective feeding practices as a result.

Nonetheless, even when feed tables are accessible, the case studies' surveys found that many farmers do not use them or use them incorrectly. The majority of tilapia farmers in China, according to researchers, did not feed their fish as per the recommended rates advised by their feed tables, and struggled to follow ambient temperature, body mass, and pond biomass into effect when setting feed rations. Unsurprisingly, the one survey farmer who kept feed records and adjusted feed rates according to the specified feed tables had the highest feed conversion ratios, demonstrating the relevance of feed management methods. Similarly, the major Indian carp farmers examined in Bangladesh did not generally monitor their feed consumption or apply FCR to assess feed efficiencies, resulting in sub-optimal feeding.

**Figure 6.8**: Monitoring feed consumption via feed trays can indicate adjustments in feed applications that maintain water quality and enhance shrimp health.

*Source: https://www.aquaculturealliance.org/advocate/feed-management-improves-profits-intensive-white-shrimp-farming/*

While the use of farm records to monitor feed consumption and efficiency varied throughout case studies, it was clear that many farmers were not maintaining appropriate manufacturing records; simple farm statistics like stocking rates, mortality, feed use, and water quality were not usually recorded. Farmers will struggle to analyze and monitor the effectiveness of their production systems and establish whether changes to their management techniques result in demonstrable increases in production efficiencies without this data (Benetti et al., 2021). Farmers must be trained in feed management procedures, encouraged to use feed tables, and guarantee that they keep adequate feed and product records. The human factor is a facet of feed management that is frequently disregarded. Feeding is often the responsibility of workers on larger shrimp farms, according to experts, who often lack technical awareness of the need for optimizing feed management, are underpaid, and are rarely rewarded to enhance feed management and efficiency.

Unsurprisingly, under these circumstances, feed management is frequently suboptimal. Indeed, it was found that hired farmworkers on several commercial tilapia farms in Malaysia tended to overfeed the fish in the incorrect notion that feeding more created higher growth rates. The same experts also pointed out that feed management schemes in shrimp production are frequently tailored to the needs of the farmer or farm worker, with little respect for the cultivated species' behavioral inclinations. Because shrimp feeding activity is strongest at night, feed regimens should concentrate on supplying feed at night, rather than during the day when it is more convenient for those providing the feed (Akter et al., 2016)

In several cases, forward-thinking farmers have reported devising their own feeding systems to maximize feed efficiency. In Andhra Pradesh, India, for example, the majority of big carp growers reported spreading farm-made food at fixed spots in their ponds. Because they simply placed their mash feeds in this manner, much of it was distributed in the water column and squandered. More forward-thinking farmers used a "bag feeding" method, in which feed combinations were deposited in bags strewn about the pond. This strategy increases demand feeding and leads to faster growth rates, increased feed absorption rates, and higher retention rates since less feed is wasted to the water column, as reported by the Indian major carp growers of Bangladesh. The invention of limited feeding regimes, in which the fish are left unfed for one day every ten days — a strategy aimed to decrease feed costs and drive compensatory growth – was another unique feed management practice described by carp farmers in Andhra Pradesh.

Some striped catfish producers in Vietnam have described a similar limited feeding technique. While restrictive feeding regimes have been proven experimentally in the North African catfish (*Clarias gariepinus*) in Africa, they have yet to be embraced as a farming strategy; carp farmers in Andhra Pradesh have also created "break feeding schedules," in which feed rations are divided into 2 rations, each delayed by 20 minutes. The dominating fish can be fed until they are full during the first feed round, while the smaller fish can be fed until they are full during the second feed round. Because the method encourages minimum size fluctuations at harvest, it enhances satiation levels across the entire culture population. The following are some additional strategies:

- the use of feeding areas to make it easier to apply floating meals and avoid waste; and
- heating high-starch mash feed materials (e.g. broken rice) to promote gelatinization, improve digestibility, and boost nutritional availability.

Clearly, the involvement of innovative farmers in improving on-farm feed management methods is critical, and systems to promote and transmit these ideas to other farmers must be developed.

## 6.7 THE ROLE OF NATURAL PRODUCTIVITY AND THE IMPLICATIONS FOR FEED MANAGEMENT

Natural productivity enhancement to offer a feed source for low-trophic feeders like tilapia, carp, and shrimp is a popular method that was widely described in case studies. Inorganic and organic fertilizers are widely employed in extensive and semi-intensive production systems; nevertheless, there are significant disparities in the types of fertilizers utilized, as well as their availability, cost, and application rates. While many farmers were able to preserve natural productivity in their farming systems, others, particularly in Africa, were found to fertilize at sub-optimal rates, resulting in poorer yields. In such instances, it would enhance the efficiency of their productions by training farmers to make use of straightforward indicators to measure natural productivity levels in their ponds and by providing data for managing their phytoplankton, zooplankton, benthos and periphyton production using proper fertilizer use. It can also be useful to improve production efficiency in large, semi-industrial systems to determine the qualitative and quantitative correlations between natural productivity and the effect on the nutrient cycling and retention in cultivation by the use of extra- and farm-made feeding stuffs (Akter et al., 2016)

The comparative function of feed in the feeding of farmed animals vs. natural production is misunderstood. Feed frequently has a double purpose in feeding the animals and in stimulating their natural productivity as a nutritional source. Acquiring a better knowledge of these dynamics is crucial if the nutrient retention of farmed species and the crop system is to be improved, feed formulation to be improved, feed costs reduced and feed management systems to be effective.

## 6.8 THE USE AND EFFICACY OF FEEDING DEVICES

One reservoir submitted a study of the feeding equipment produced for salmonids and used more and more in other species. The case studies showed that small-level farmers rely heavily on hand feeding; automated feeding equipment only have been included in selected areas. For example, while the majority of tilapia growers in China are using automated feeding system, only selected farms in Thailand and Malaysia have been reported and not in place in the Philippines, Ghana and much of sub-Saharan Africa, which still relies on hand-feeding for the vast bulk of SME farmers.

**Figure 6.9**: Auto-feeders intermittently distribute small quantities of feed that are caught by the shrimp before sinking to the pond bottom and losing nutrients.

*Source: https://www.aquaculturealliance.org/advocate/feed-management-improves-profits-intensive-white-shrimp-farming/*

The scenario with Indian major carp producers has been similar, although farmers also choose to rely on manual feeding, except for certain feeding tactics like bag feeding technology, which effectively is a simple kind of request feeding. The manufacturing of shrimp is dependent mostly on hand feeding and the use of feeding tables to check consumption. While feeding by hand offers farmers the advantage of monitoring feeding performance and of adapting rationing accordingly, automated feeders can be cost-efficient, decrease demands on work and enable huge volumes of feed to be supplied efficiently.

When it comes to demand feeders, they have the benefit of taking the behaviors of farmers into account, the retrogrades of hunger and the nutrient quality of the food. They also have the advantage over handfeeding. Depending on the farmed species, fish size and the cropping system design the choice and intricacy of the automated systems are necessary. For instance, in hatchers, simple belt feeders can be used to give minimal amounts of food to be fried, often requiring almost continual feeding.

More complicated systems, like static demand feeders and moving mechanical systems based on pressurized gas, can be employed in grow-out systems. Computer-controlled automatic feeders, employing video or infrared sensors to monitor intake, can be especially useful in species where feeding hierarchies emerge, to ensure that all fish are satiated. For instance, while dominant Atlantic salmon graze on the top and quickly become satiated, the subdominant fish eat less in the water column and require more time to reach satisfaction. Connecting feed control systems to automated feeders guarantees the satiation of dominant and sub-dominant fish.

While several of the more complicated processes require power and rather high technical know-how, installations and operations would also be expensive, which would be inappropriate for small-scale rural farmers. Alternatives exist, though, such as simple mechanical demand feeders, to enhance feed management on-farm by smallholder farmers. The use and effectiveness of these mechanisms must be created and encouraged where necessary.

**Figure 6.10**: In biofloc systems, strong aeration keeps organic matter in perpetual suspension. Significant amounts of microbiota "piggyback" on the organic matter, recycling the rich nutrients available in the water.

*Source: https://www.aquaculturealliance.org/advocate/feed-management-improves-profits-intensive-white-shrimp-farming/*

## 6.9 FEED MANAGEMENT AND THE ENVIRONMENT

Feed quality and feed management procedures play an essential role. Food methods that are inappropriate and in particular lead to overfeeding are a significant source of extra nutrients into the environment. Overfeeding generates less feed efficiency, more feed waste and more environmental deterioration in many circumstances. About 15–25 percent of the nitrogen in feed is retained by culture-animals, the remaining ones entering the environment in soluble or particulate form, according to the effectiveness of feed composition and feed management procedures (Waagbo & Remo, 2020).

More than 75% of the total carbon and phosphorus in the feed is excreted or released as particulates or through the gills. The negative environmental effects of nutrient and organic enrichment include increased demand for biochemical oxygen in the waterbodies and sediments (anoxic sediment), alterations in the structures of the community and eutrophication. The optimization of feed formulations, quality management procedures and feed management can play a vital role in restricting nutrient inputs to water and reducing the environmental implications of farming activities.

Increased nutrient charges and an increased ability to damage the receiving environment are seen in the feed conversion rate connected to poor feed and/or feed management. There were considerable changes in FCR, both feed type and feed management, in the case studies. For instance, when commercially produced and farm-made feeds have been employed in semi-intensive strapping catfish production in Vietnam, experts recorded an FCR (food conversion ratio) of 1.6:1 and 2.9:1. Although there would be differences between the two feed types in nutritional composition and management procedures, the findings indicate that commercially produced pellets were efficient than agricultural-generated feeds, and the industrially processed feed likely resulted in lower effluent loads. Similarly, in India, Ramakrishna, an Indian semi-intensive pond culture, FCR's 1.8–3.4:1 and 2.3–4.1-1, respectively, were recorded by a further set of researchers utilizing commodity-produced pellets and farmed feed (Tucker & Hargreaves, 2009).

Despite the differences in feed formulas and the procedures used in feed management, it may reasonably be suggested that feeds produced less pollutants overall. The quality, the accessibility (temporal and spatial) and the high cost of many food products used in farm feeds were recurrent themes in the case studies. The quality of farmed foods often depends on the availability of ingredients, costs and farmers' judgments of the right amount of food to be used. As nutritional imbalances contribute to lower productivity and an increase in effluents, the on-farm feed formulas used by farmers often need to be improved.

The use of cost-effective formulas which are waterproof, palatable, have a nutrient content aimed at a particular developmental stage of fish and, where applicable, will improve productivity parameters (growing and FCR) by taking into account the endogenous availability of organisms of nature such as the cultivation of shrimp, semi-intensive Indian major carps, tilapia culture). Similarly, inadequate feed handling, storage and rotting before ingestion will lead to low feed effectiveness and increasing damage of the environment.

In many case studies provided, important interventions were sub-optimal feed management methods and the need to create and promote improved feed management practices. Improvement of feed management procedures would increase the economy of production and lessen the impact on the environment of farms.

## 6.10 THE ECONOMICS OF FEED MANAGEMENT

The effects on the economy of the farming business of feed type, formulation and feed management procedures are essential matters which farmers must take into account when planning agricultural activities. While farmers typically have difficulty assessing these economic links, they can have a tremendous effect on the profitability of the farming activity. These correlations are particularly visible in the highly competitive industries, with feed costs accounting for a large percentage of production costs, low farmer's prices and marginal profitability,

**Figure 6.11**: Economic planning is crucial.

*Source: https://www.aller-aqua.com/press/articles/economy-in-fish-farming*

One researcher showed that feed expenditures accounted for 82.9 percent of the overall production costs of striped catfish producers in Vietnam using feed produced commercially. When a combination of farmed feed and commercially produced feed were employed, this figure was lowered to 79.0% and further reduced to 77.4% when the only food source was farm-made feed.

Although the use of farmed feed looks to be the cheaper alternative and a change to them cuts investment costs, they are less effective in growth and FCR terms, which means that they are more expensive to apply in terms of real cost of production (cost/kg of fish produced). A study showed a combination of commercially manufactured and farm-produced feeds with US$0.88/kg fish, compared to US$0.79/kg fish for farmers that used commodities.

There is a higher income of $0.09/kg compared to the exclusive usage of agriculture feedingstuffs when feeding commercial feedstuffs or combined with agricultural feedingstuffs. This implies a substantial cost save - a key element when farms have so low prices. The expert added that the high costs of commercially produced pellets forced poor farmers to return to agricultural feed or utilize it when the prices of farms were low. The economic assessment reveals that while returning to farm food will cut feed prices, farmers must know that there will also be a reduction in profits whilst there is some margin for substitution with a feeder (Tacon&Halwart, 2007).

Obviously, low-income farmers use less expensive, but eventually more productive and commercially produced feed to make the most lucrative use. In various industries, financial arrangements have been formed between feed producers, distributors and farmers. These forms of micro-loans should be encouraged and innovations should be found to finance feed procurements, perhaps by the establishment of organizations and associations of farmers, bulk buying arrangements, the participation of banks and micro-loans, and the establishment of public-private partnerships.

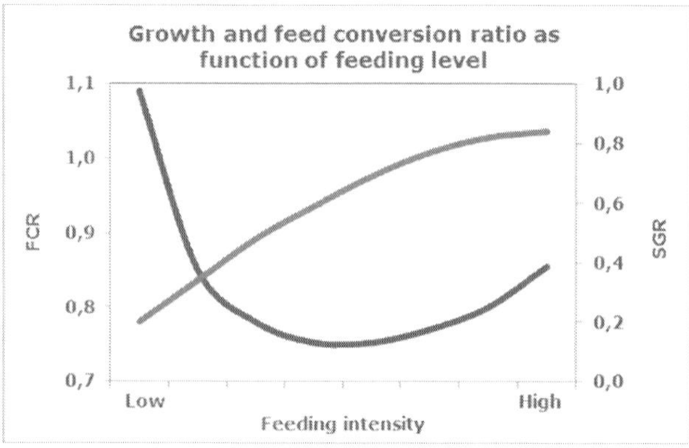


**Figure 6.12**: The correlation between feeding intensity, growth and FCR.

*Source: https://www.aller-aqua.com/press/articles/economy-in-fish-farming*

The link between the management of feed and the economic effectiveness of the agricultural operation is of importance to farmers and justifies choosing one feed management method over another. This is significant for farmers. Studies have shown how decreasing feed management techniques, as seen as a drop of FCRs, affected the economic sustainability of a farm operation, based on a bioeconomic model for an intensive recirculating system to cultivate Japanese meagre (*Argylosomus japonicus*).

For example, on an annual farm of 600 tonnes, an increase of FCR from 1.0:1 to 1.6:1 has raised feed prices by 36.2% to 46.1% and an increase of 22.9 percent over the yearly cost of farms. The investigation indicated that optimizing the feed management strategies that influence FCR and eventually profitability are economically important. These include the optimization of feeding frequency, ration and raising the temperature, among other things. In Thailand's tilapia culture, researchers have also evaluated feed utilization and its effect on growing times and costs for production and have found that additional considerations, like excavation costs of land, water and ponds, must be explored as well as the above-mentioned factors. In circumstances where land rental or purchase costs were costly, larger density fed-lot systems would become more cost-effective.

A wide number of characteristics that affect the use and eventually profitability of on-farm feed must be taken into account by farmers. Main parameters include: system design (e.g. extensive versus intense agriculture systems); operating parameters (e.g. temperature; water quality) (e.g. feeding schedules). Farmers do not generally have access to this information, therefore economic tools must be developed to help farmers comprehend the consequences of their feed choices and management tactics.

## 6.11 AQUAFEED POLICY, REGULATIONS AND GOVERNANCE

In several of these cases, poor regulatory control and lack of standards were identified as limits on feed supply, quality and usage across the whole waterfeed value chain. Major reported problems were: the absence of adequate labeling for products; and/or the absence of standard feed specification, which resulted in inadequate formulations being marketed to farmers. Since the quality of feed supplied has been limited, farmers mainly rely on feed ingredient suppliers and producers to supply excellent products.

This problem is particularly relevant to poor resource farmers, who have limited practical alternatives to the provider or manufacturer to buy poor quality or under-standard feed and feeding stuffs. In the light of these issues, it is relevant to ensure the quality of feed and food ingredients in aquafeed regulation and governance and to optimize the production of feed. An analysis of the governance structures and the role played by the legal, policy and regulation tools to ensure feed quality indicated that the regulatory instruments used to oversee the sector have considerable regional differences.

In industrialized nations, in which the industry is defined by large-scale industrial production and feed makers, it is obvious that the quality of commercial feed is generally good (a more consolidated feed manufacturing sector). In these nations, the feed production, use and management rules are aimed to ensure the quality of feed and to limit its influence on the environment. Conversely, numerous small-scale farmers, employing farmed feedingstuffs or commercially produced feeds obtained from a pool of small feed producers, dominate in less industrialized regions like Asia (Subasinghe, 2017).

It is possible that laws in these places are more focused on regulating feed quality, ensuring that farmers have access to better feed, with less emphasis on the effect of its use on the environment. In actuality, few Asian nations have regulatory mechanisms in place to regularly monitor ingredients and feed quality; in other countries like as sub-Saharan Africa, regulatory actions are surely not available. Nevertheless, the development and adoption in less developed producer countries of proper regulatory frameworks for feed is positive. For instance, new regulatory framework for the manufacture and trade of aquafeeds was recently created in Bangladesh and Vietnam.

These new laws involve the registration of feed factories, the use of feed regulations and the continuous monitoring of products in Vietnam. However, no feed legislation for freshwater aquaculture is currently in effect in India. The Coastal Aquaculture Authority (CAA) has limited authority to revise the registration of feed mills in salt and brackish-water systems and is now subject to voluntary rules of practice in the aquafeed producing industry. The CAA plans to implement a legal framework in the near future.

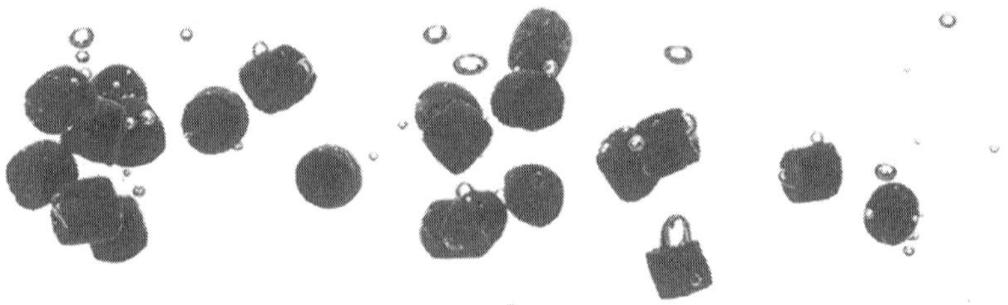

**Figure 6.13**: Fish food pellets.

*Source: https://www.aller-aqua.com/press/articles/economy-in-fish-farming*

No regulation of feed composition, feed performance, feed use, treatment of effluents or levels, nor development or adoption of rules of conduct or better management practices are seen in sub-Saharan Africa.

Government institutions traditionally offer regulatory frameworks, policies and regulations that restrict the use of aquaculture and aquafeed. In recent years, a new approach to governance has seen the creation of certifying bodies like as the Aquaculture Stewardship Council (ASC), the Aquaculture Certification Council (ACC).

In many ways, these "non-state market-driven systems" now contend in what some authors have called the "privatization of governance." Market access is steadily being driven by certificate programs and conformity standards, especially export-oriented markets into the industrialized world. Feed and feeding are significant parts of those requirements and from a market access point of view, regulatory action to control and monitor feeds and feed ingredients are becoming increasingly important for governments to assure compliance.

In addition to developing legal frameworks to control the quality of feed that farmers have, institutional capacity to comply with the requirements needs to be developed. The quality and manufacture of feed is a complex and expensive activity and case studies reveal the absence of institutional capacity for its implementation in many countries with developments in the right legislative framework. There is also a need to explore the role that NGOs such as national organizations in the feed industry can play in supporting the sustainable development of aquafeed production.

## 6.12 TRAINING AND DEVELOPING INFORMATION NETWORKS TO IMPROVE FEED MANAGEMENT

It is important to address the issue of the training and dissemination of data to farmers and especially small-scale farmers who are less exposed to the latest development technology and management. The majority of farmers who utilize farm feedingstuffs have minimal household training and are therefore troublesome in the transmission of sophisticated technical signals, requiring constant attention. Weak networks of extension and dissemination of knowledge lead to low adoption and improved management practices of new feed technology.

Farmers' clusters and associations have demonstrated their effectiveness in disseminating information and increasing farmer training. The clustering of farmers is growing throughout Asia, and in several nations in Africa more recently. In Kenya, for example, smallholders in India, Indonesia, Bangladesh, and Vietnam are now being pushed to establish farmers' associations and in the whole of Asia, smallholders are being organized more and less into groups/clusters.

Furthermore, it has shown beneficial to identify and train essential, innovative farmers for the training and further promotion of other farmers and farmers' camps. While training requirements are sector-specific and more detailed in case studies, it is clear that they usually focus on improving feed formulations; formulating species- and lifestyle diets; and improving the understanding of quality ingredients, nutrients and selection, manufacturing, storage and management procedures for on-farm feed. Access for small-scale feed producers and farmers manufacturing agricultural feed, to the latest market information is a problem that needs to be tackled (Anderson et al., 2019).

While large-scale production activities are generally accessible to the farmers and small-scale feed producers and are aware of the cost, accessibility and nutrient content of their components on the market. The development of economically efficient farm feeds requires the establishment of contemporary market knowledge, including sources, suppliers, quality and pricing. In addition, it should also be encouraged to use appropriate feed items that are accessible locally and seasonally and can be incorporated into agricultural foods.

The availability of such sources of ingredients and how they might best be included into their formulations must be made known to growers and small-scale feed producers. At present, there is a need to encourage programs that employ local media to provide farmers with updated feed component availability, quality, prices and supplier information. In addition, the information networks are not effective or lacking. Furthermore, farmers need access to information on feed formulations and inclusion rates for species-specific ingredients. Areas-specific feed component databases and cost information that farmers and small feed companies can easily access could also be considered.

Fish agriculture is intensified when a higher demand for farmed fish, i.e. near towns and cities, is present. It is not intended to appeal or is a result of development efforts to the supporters of fish farming. This is the product of a successful market. The sustainable development of aquaculture depends on fish nutrition, feed, and feed management. An aquaculturer has a key role in selecting a proper feed. The long-term sustainability of aquaculture is ensured by a feeding strategy that uses cost-effective low-fish meals and low pollution water feeds in an extra feeding system. The aquaculturalist is also responsible for applying sound feeding practices to boost fish productivity, profitability and prevent cultural deterioration.

# Chapter 7 ROLE OF THE GOVERNMENT IN INTENSIVE AQUACULTURE

## CONTENTS


7.1 Introduction.......... 162

7.2 The History of Aquaculture Regulatory Framework .......... 164

7.3 Establishing an Institutional Framework .......... 167

7.4 Licensing Structures.......... 169

7.5 Access to Land .......... 170

7.6 Access to Water Resources .......... 172

7.7 Aquaculture Environmental Management.......... 172

7.8 Site Selection and Environmental Impact Assessment (EIA) .......... 174

7.9 Effluent Regulation .......... 175

7.10 Waste and Waste Management.......... 177

7.11 Coastal Aquaculture as Part of Integrated Coastal Management .......... 177

7.12 Escapement.......... 179

7.13 Regulations of Trans-Boundary Aquaculture Ecosystems .......... 180

7.14 Trade and Health Issues .......... 180

7.15 Introduction and Movement of Species.......... 181

7.16 Food Safety and Health.......... 181

7.17 Mitigating Investment Risks .......... 182

7.18 Aquaculture and Zoning.......... 183

7.19 Simplification of Administrative Processes: Single Window Approach... 185

7.20 Financial Incentives.......... 186

7.21 Social Impacts.......... 187

7.22 Conclusion .......... 188

Aquaculture for countries, particularly in underdeveloped countries, is quickly increasing as a source of food and money. The regulation of the aquatic ecosystem, the environment and general aquatic health industry must be founded on a robust regulatory framework governed by stable institutional arrangements. Many aspects need to be taken into consideration. In this context, public sector institutions are especially significant and should meticulously identify national aquaculture sector aims and objectives. Institutional underpinnings and core legislative requirements are designed to help and safeguard aquaculture farmers, attract investments in the state ensuring regional and international exports and consumption standards are maintained and generally protect the environment. The aim of this chapter is to present the main challenges to be addressed when building a national regulatory framework for aquaculture. Moreover, the newest trends and advances in the legislation on aquaculture in the last 20 years and the most common issues are given.

**Figure 7.1**: Aquaculture farm.

*Source: https://www.conserve-energy-future.com/aquaculture-types-benefits-importance.php*

## 7.1 INTRODUCTION

A variety of important institutional and technological requirements depend on the ability to operate sustainable aquaculture activities. In each step of the development of an aquaculture program, entities of the public sector play a significant role. Several concerns have to be tackled to ensure that all aquaculture operations that take place while reducing risk and effects have their full potential. Appropriate planning, zoning and priority setting for the various prospective consumers are the basics for a sustainable aquaculture business. Disputes with other participants and business factors (e.g., fisheries, agriculture, forestry, tourism, and subsistence users) can be costly and result in investment failure as well as irreparable damage to environmentally and economically essential resources (Stentiford et al., 2020).

The unambiguous identification of national policies and practices is another important component. Before the development of the industry explosions, the governments should create clear policies and processes. These policies and procedures must not be comprehensive but should address the most prevalent problems and implications involving aquaculture. This implies that the process for obtaining permits, licensing, concessions, property ownership, and project approvals should be straightforward and transparent. This is one method of ensuring that all the players follow the proper procedures and that government authorities receive clear information on how to continue with approvals.

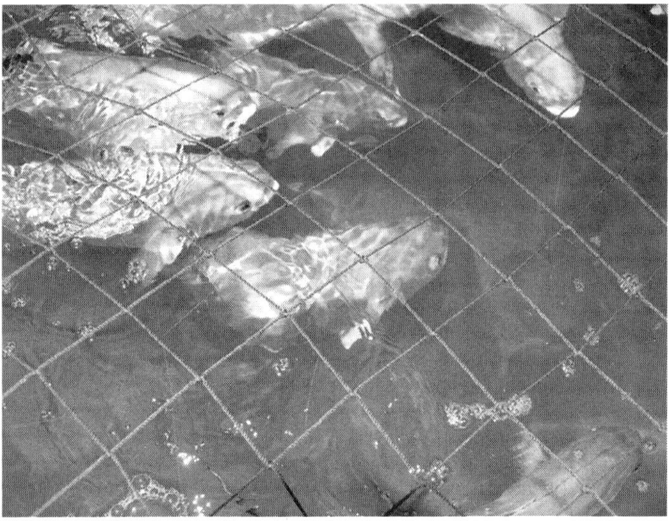

**Figure 7.2**: Will the U.S. development of aquaculture technology for emerging species like pompano lead to commercial success outside the United States?

*Source: https://www.aquaculturealliance.org/advocate/government-policies-key-to-aquaculture-development/*

A solid legislative framework should be put in place to defend the rights of aquaculture producers while also protecting the interests of the state. This framework should include minimal standards for the aforementioned procedures and institutional architecture. Bans, limits, land-use categorization and zoning, environmental impact assessment, mitigation plans, permits, user fees, performance bonds, and monitoring criteria are some more measures for controlling aquaculture activities. Environmental impact assessments, effluent outflow licenses, regulations on the use of non-native species, restrictions on medicine and chemical usage, feed content requirements, feed use regulations, and other management measures are all commonly used to regulate aquaculture.

## 7.2 THE HISTORY OF AQUACULTURE REGULATORY FRAMEWORK

Countries that have experienced some level of aquaculture development have established a legal framework that allows for control over access to and operation of aquaculture activities. The legislative framework also establishes the legal structure and directions for aquaculture quality evaluation, including formalities connected to certain processes.

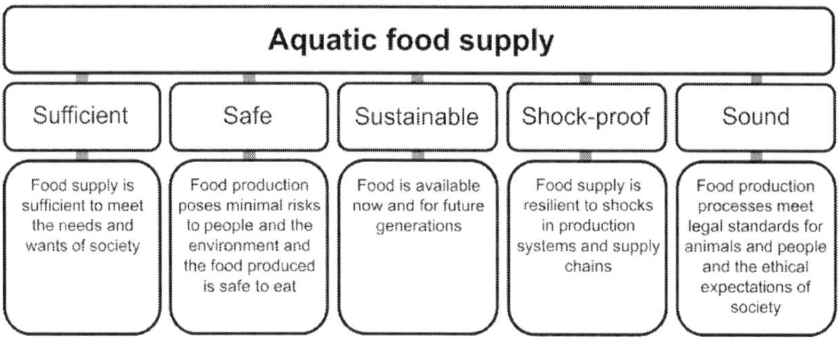


**Figure 7.3**: Aquatic food security.

*Source: https://onlinelibrary.wiley.com/doi/full/10.1111/faf.12152*

Other sectoral laws have a direct impact on aquaculture, such as land law, which includes the use of public spaces like mangrove areas, water law, environmental legislation, animal welfare and animal disease law, fishery law, and trade regulations. Many of the challenges and concerns raised are not exclusive to aquaculture and may be addressed under a broader legislative framework. Furthermore, many of the existing laws and regulations may not even apply specifically to aquaculture and are thus frequently implemented inconsistently to the sector. Disputes may emerge within the scope of aquaculture legislation or among the authorities and institutions involved.

Several countries have implemented aquaculture-specific legislation, basic fisheries law, water law, or another type of legislation. These laws often establish some aquaculture concepts and then entrust the legitimate authority with the right to govern aquaculture.

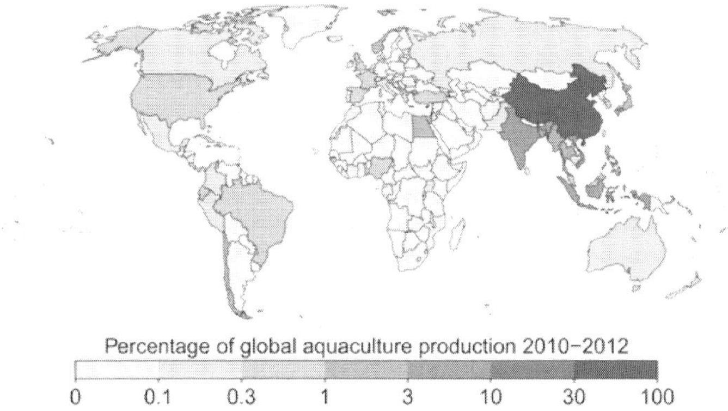


**Figure 7.4**: Proportion of global aquaculture production by volume by country. Data from FAO (2015).

*Source: https://onlinelibrary.wiley.com/doi/full/10.1111/faf.12152*

It has become increasingly usual to govern capture fisheries and aquaculture in the same piece of legislation, despite the fact that aquaculture is more closely related to agriculture than capture fisheries. Governments in nations where the aquaculture sector is developing find it helpful to have the same body enforce and supervise both sectors. Nonetheless, the two should be legally and institutionally segregated. From a legal standpoint, existing fisheries regulation is frequently insufficient for controlling aquaculture.

Modifying legislation is frequently a more involved procedure than establishing and implementing a new one. Many aspects must be addressed for aquaculture that are not naturally addressed under fisheries legislation, such as nutrient supply, disease monitoring and control, the use of medications and pesticides, temperature management, and water quality. Individual legislation is thus the finest and most sound way to regulate aquaculture (Rimmer et al., 2013).

Many of the recently enacted fisheries laws address certain elements of aquaculture. In other situations, the fisheries and aquaculture industries are even listed together, showing that the legislature values them equally Madagascar's Fisheries and Aquaculture Ordinance of 1993, Burkina Faso's Forestry Code of 1997, Malawi's Fisheries Act of 1997, Zambia's Fisheries Act 2011, and Chad's Forestry, Fauna, and Aquatic Resources Act 2008 are some examples.

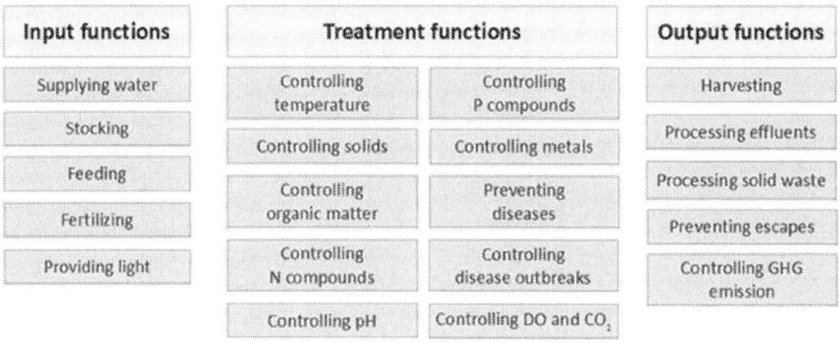


**Figure 7.5**: Simplified aquaculture production functions map.

*Source: https://www.researchgate.net/publication/339924494_A_Definition_of_Aquaculture_Intensity_Based_on_Production_Functions-The_Aquaculture_Production_Intensity_Scale_APIS/figures*

The development of the aquaculture business has resulted in severe environmental harm, and governments are increasingly realizing that unpredictable and ineffective legislative structures are seriously impeding the sector's long-term development. In terms of methodology and the required basic elements for sustainable management aquaculture rules and regulations have established some consistency. This is being driven by a number of factors, including: increased political awareness as the economic importance and prospects of aquaculture become more apparent; increased awareness that unsuitable laws and institutional arrangements can significantly impede the sector's development; and evidence of ecological harm and social disruption as a result of the emergence and poorly regulated expansion of the sector. The following research illustrates the most notable trends and actions in recent years in national aquaculture legislation, as well as the most common aspect found in aquaculture law.

We are now widely aware of the need to adopt policies and a legal framework for aquaculture. Legal and institutional instruments must be developed to:

- Aquaculture is recognized as a separate agriculture sector;
- Integration in resource use and development planning of aquaculture problems;
- Improving food safety and quality to protect consumers and to fulfill importer criteria; and
- Improving aquaculture management, especially when it has a social or environmental unsustainability risk.

Things such as registration and admission, strategic planning, the many environmental effects of fisheries should be addressed consistently, so that industry, the ecosystem, other users of resources and consumers are protected. They must be addressed. States that have become farm-shrimp producers, such as the Philippines, have acknowledged the necessity of integrating these features in new regulatory frameworks. The 1998 Philippine Fishery Code provides for fishing and aquatic resources development, management and conservation and includes a variety of issues, such as land use, fish leasing, institutionalization of standards of practice for fish farming, incentive schemes and disincentivizes for sustainable aquaculture practices, and the creation of Aquaculture Investment Fund (Rahman, 2014).

## 7.3 ESTABLISHING AN INSTITUTIONAL FRAMEWORK

The national governments have the most significant responsibility in planning and managing the sustainable spread of aquaculture. The government's sole duty is to define the national aquaculture development agenda, including the appropriate procedures for administrative, regulatory and enforcement measures. Aquaculture policy has become increasingly prevalent as a whole in the public sector. Governments that aim to promote sustainability should adopt clear and feasible, financially, socially and environmentally sustainable aquacultural development policies. In addition, sustainable policies should be developed by the dialog between national authorities, NGOs, the commercial sector and international organizations.

There are several institutional duties for the necessity for the Member States, as stated in Article 9.1.1 of the FAO 1995 Responsible Fisheries Code of Conduct (CCRF), to create, promote and improve an acceptable legal and administering structure for aquaculture development. Legislation is only as effective as implementation and enforcement, which could be catastrophic to the execution of the policy goals to be ensured by regulation. Aquaculture has a nature that requires governments to undertake a range of broad oversight roles in terms of general policy and industry performance (Roubach et al., 2019).

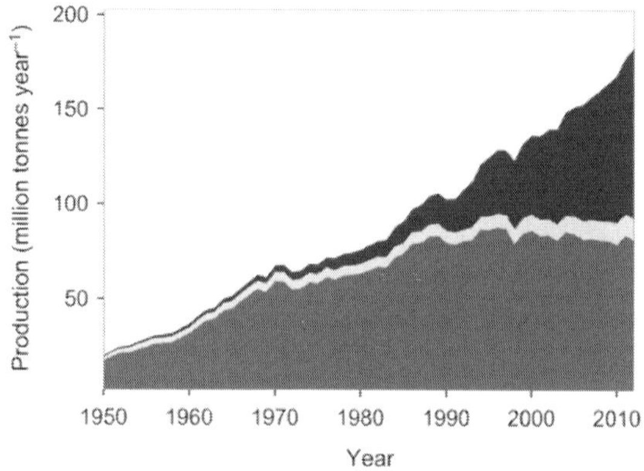


**Figure 7.6**: Trends in global production from marine (red)- and freshwater (yellow)-capture fisheries and aquaculture (blue). Data from FAO FishStatJ (FAO 2015).

*Source:* https://onlinelibrary.wiley.com/doi/full/10.1111/faf.12152

The central Government must have the obligation to establish a coordinated overall approach for the aquaculture sector, issues of national interest such as the design of a comprehensive federal policy. Likewise, it is up to the Government and central law to adopt suitable legislation governing aquaculture.

Nevertheless, this does not prevent the enactment at national, regional and municipal levels of implementing regulations. The overarching pattern of administrative duty assignment involves the federal government's allocation of full responsibility for aquaculture policies and regulations.

This role, however, is allocated to several levels among the local authorities in federal States, each of whom is responsible for local and technical issues.

Administrative overlapping and interference are the most typical problems in the management of aquaculture policy. Efforts are doubled and disagreement between departments and the levels of government are common. Laws on industrial aquaculture development: the regulatory framework for aquaculture must ensure the operator has a safe right to perform aquaculture activities on land on the fish farm and on good quality water. It also has to control and plan the competing use of resources such as land and water by means of a system of authorization or licenses. Licenses help protect aquaculture's environmental sustainability by analyzing the appropriateness of proposed fish farm sites and the environmental consequences of their activities.

## 7.4 LICENSING STRUCTURES

The first step in the oversight of aquaculture operations is to compel an operator before carrying out aquaculture activities to acquire an authorization or a permit. The title of the document is not important, but the term chosen depends upon its widespread usage or customs in each country, whether it is referred to as an authorization, permit, or license. The Ukrainian legislature, for example, was reluctant to use the word license when drafting the 2012 Aquaculture Law since it contradicted its existing license law (Rodrigues et al., 2012).

However, two fundamental functions are important for the text. To begin with, the permit must provide the government authorities with a legal foundation for monitoring and monitoring all aquaculture operations. This enables the government to evaluate the environmental sustainability of the plan and to set criteria for the sustainable operation of the farm. Secondly, the authorization must also provide a clear right for farmers, provided that the operator adheres with the conditions of the license, environmental legislation and any equivalent aquaculture practice guidelines.

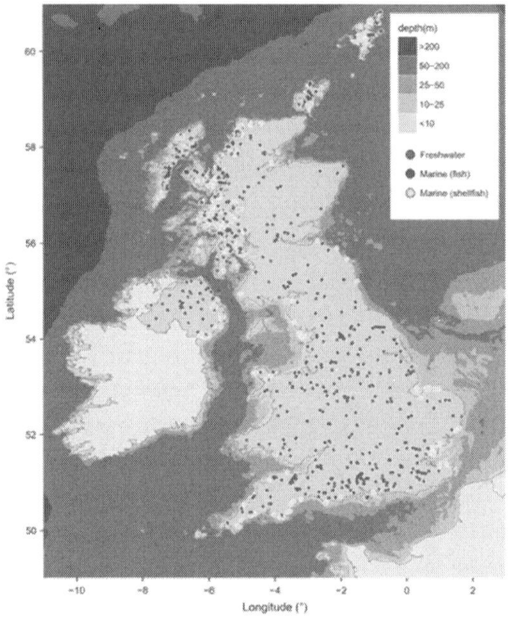


**Figure 7.7**: Location of aquaculture sites in the UK. Location and farm-type data from information compiled in the UK Multiannual plan for the development of sustainable aquaculture.

*Source: https://onlinelibrary.wiley.com/doi/full/10.1111/faf.12152*

A license may be an absolute necessity for an aquaculture facility to be established or may impose environmental or other requirements. Allowable systems can be universal or may be available just for farms of specified sizes. The license may also be subject to a cost. "Licensing for Development" is a generic useful word to define any system of regulation of public land use that requires landowners to provide specific licenses to develop or alter land-use systems for certain objectives. This is the imposition of a sanction by illicit development of private or public lands without the required license. Countries that have put into practice some sort of authorized licensing structure include Canada, Norway, Madagascar, Zambia, Australia, Cambodia, Gabon and the Democratic People's Republic of Korea.

## 7.5 ACCESS TO LAND

Aquaculture purchases depend on the overall structure of land ownership and land holdings in each national authority. The difference here exists between countries which allow private ownership of property and those which control land ownership. One of the important issues is whether environmental and social requirements are imposed on landowners by the acquisition of land rights for fish farming.

Because the farmers are placed on any land, they must have a safe legal right. This is mostly not a big concern for freshwater aquaculture, as land regulations typically enable farmers to gain ownership interests. Special provisions may nonetheless apply if the land belongs to the state, to the coastal zone, or to the customary tenure system (Rosa et al., 2012).

There is added issue, however, of whether rights to use of the body of the water body sub- and adjoining to it must be provided in some nations where land rights are under total State control. These countries, often situated in the regions of Central Asia and the Caucasus, refer not to an aquaculture installation, but to land leasing in their law.

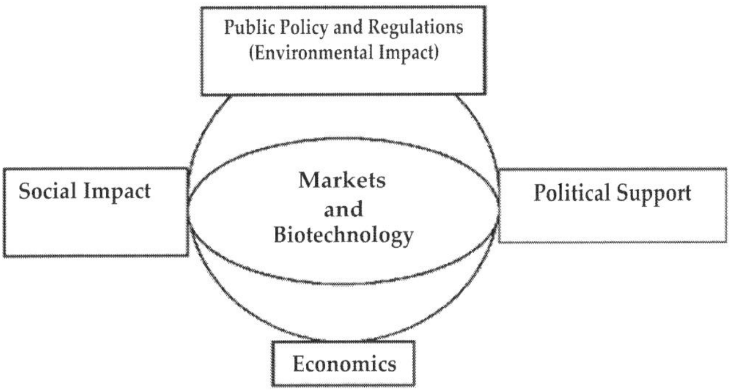


**Figure 7.8**: Promote sustainable commercial aquaculture.

*Source: http://www.fao.org/3/X9894E/x9894e05.htm*

Fish farming can occasionally present a competitive problem for other resource-use farming enterprises. Fish agriculture involves soil, water, work, time and inputs of nutrients. It must therefore be integrated into the current agricultural system. The integration of many farms can boost the efficiency of the farm as a whole and hence increase opportunities for adoption.

Integration can also include a complementary activities and ensure that resources be used more effectively. The best opportunity for success is the improvement of efficient use of those resources that limit overall agricultural production. For instance, the quantity of water available can limit expanded production and improve the efficiency of use of water can enhance total agricultural productivity. The goal of integrating aquaculture and agriculture is to boost overall productivity by improving water use, work and diversification of agricultural produce (Rico et al., 2017).

The integration of farming into other uses, such as aquaculture or inland fishing, can ultimately attain resource-use efficiency. Greater awareness of the prospects and financial advantages of such integration would allow agricultural policymakers and planners to come into actuality. With increasing pressures on resources, this drive for integration is likely to intensify. Research that generates new types of integrated water-farming systems and other creative systems that successfully respond to environmental and resource concerns also needs to be encouraged and promoted.

## 7.6 ACCESS TO WATER RESOURCES

While proper water quality supplies are important for any aquaculture plant, this need can also lead to disputes with other water users seeking the same water for other uses such as farming, fishing, industry, leisure activities or water supply. Due to this rivalry, access to water may be quite difficult for fish farms. Accessibility to the seaside can be a huge challenge in developed countries. Water rights may be issued, including limiting water amounts which may be removed under a certain licence, subject to various limitations. The overall trend is for States with concessions or environmental licenses to restrict their water consumption.

## 7.7 AQUACULTURE ENVIRONMENTAL MANAGEMENT

Uncontrolled development of aquaculture can harm the environment, disturb rural people and promote excessive use of natural resources. Environmental management aims at minimizing, preventing or mitigating negative environmental effects of human activities to allow improved practices to allow more sustainable use of resources. To establish whether environmental management is effective, the following conditions may be helpful:

- identifies the potential detrimental effects on the environment of human activity;
- rules to specify changes in allowed variables without unacceptable environmental impacts must be drawn up;

- Management approaches technology-based to prevent excessive alterations in ambient variables must be made available.

The conventional obligation of safeguarding environmental quality is with governments and the business sector. Government is responsible for setting environmental management standards and enforcing compliance with them through the adoption of rules and legislation. While the private sector is responsible for the expense of assessments in order to establish probable ecological repercussions, government guidelines and recommendations of such assessments must be provided for.

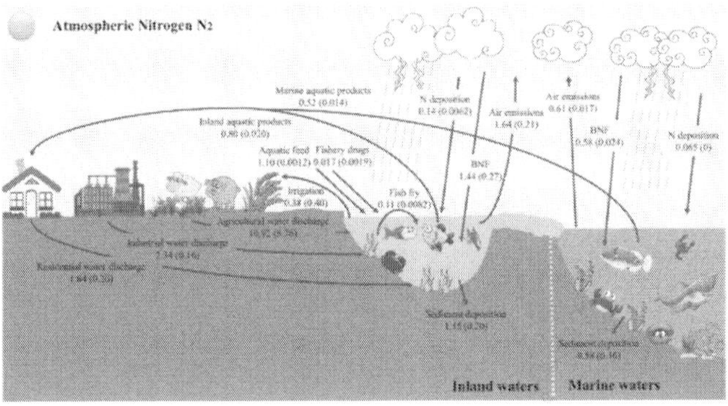


**Figure 7.9**: Schematic model of national Nr cycling in China's aquaculture ecosystem.

*Source: https://www.nature.com/articles/s41598-018-29214-y*

The development of aquaculture activities gives rise to growing environmental concerns. Management shortcomings or accidents can have serious negative effects on aquaculture operations. The conventional obligation of safeguarding environmental quality is with governments and the business sector. Government is responsible for setting environmental management standards and enforcing compliance with them through the adoption of rules and legislation. While the private sector is responsible for the expense of assessments in order to establish probable ecological repercussions, government guidelines and recommendations of such assessments must be provided for (Raja & Jithendran, 2015).

The Aquaculture Ecosystem approach is a method that receives increasing attention in aquaculture management (EAA). The EAA is intended to promote and manage the aquaculturing sector in such a way that the broader ecological system is not threatened by the sustainable development of interconnected ecosystems. In order to achieve these objectives, environmental issues are highly crucial.

In the policy framework of a country, the ecosystem approach should largely be emphasized. Legislation can be made in numerous ways under an ecosystem approach and is commonly done in a number of ways. In essence, the EAA can be seen as an overarching concept and is thus included in an aquaculture act's opening provisions. EAA law also contains provisions to enhance institutions and related management solutions.

## 7.8 SITE SELECTION AND ENVIRONMENTAL IMPACT ASSESSMENT (EIA)

The placement of farms at lower or marginal locations is one of the main reasons for the failure of fish farms or for detrimental environmental impacts. 16 An extensive site assessment should therefore be carried out to establish whether site attributes are appropriate to construct and operate on a farm sustainably. These assessments should identify water availability and quality, weather, freshwater flow and associated issues. In a detailed environmental impact assessment, the information collected should also be included (EIA).

Application of a general aquaculture license is normally a prerequisite of an EIA. The EIA tackles different elements, including the anticipated size and the sensitivity of the affected region of the aquaculture farm. Due to the possible disastrous position of an aquacultural farm on an unwanted site, the suitability of the site should be addressed expressly by the government before permission is granted. This judgment also provides for the incorporation of plans for coastal and lakefront maintenance by government bodies into aquaculture farm sites. In some situations, the prospective aquaculture plant's planned site and other environmental implications of the proposed concept should be considered. Therefore, when an EIA is necessary is vital to determine (Quesada et al., 2013).

If a license is submitted, the first judgment must be decided as regards the subject-matter of the project under an EIA. The legislation should explicitly establish criteria to determine if the evaluation for a proposed aquaculture project is required. As an EIA might charge the applicant a high cost, it should only be necessary for projects that pose a substantial risk of damage to the environment.

Factions such as operation size site capacity to be maximized, potential waste dumping in sensitive areas, protecting exotic species or products from modern technology or GMOs and whether rare or endangered species are at risk should be the basis for evaluating if this risk exists. Criteria for the proposal should include whether the project exists. When an EIA is needed, the decision by the authority to grant a permit should reflect the conclusions of the assessment, as should the conditions of the authorization itself. If no EIA is needed, licenses based on expressly specified guidelines should be issued by the regulator. "It is important for particular judgments on development grants or retainment to be made on the basis of explicit policy concepts in order to prepare aquaculture legislation.

## 7.9 EFFLUENT REGULATION

One major problem of operations at aquaculture farms is their ability to produce large quantities of waste products that can have serious adverse effects on the environment if not adequately managed. Waste items can affect the aquatic ecosystem and provide a health risk. The regulation of insufficient effluent from aquaculture has two different regulatory techniques. Firstly, a discharge licensing system will be imposed that will establish allowable limits of pollutants in wastewater from fish farms and related penalties for exceeding such quantities. The second strategy is banning or limiting the use of special medications and chemicals known to damage the quality of water and ecosystems. Discharge licensing procedures including water quality requirements and/or water volume require regular monitoring to certify compliance at certain frequencies. This type of release authorization is ideal for major operations with one or more effluent waste.

The most effective approach seems to use general effluent permits, which mandate farmers to apply a set of best management (BMPs) operating practices, with minimum criteria for basic monitoring. Control of medication and chemical use In the aquacultural business, overuse and misuse of chemicals pollutes and contaminates the aquatic environment. This is particularly problematic when chemicals previously designed for other industries, especially agriculture, are adopted by the booming aquaculture industry. Moreover, aquaculture farms are particularly vulnerable and can have a negative impact on the quality of aquaculture products, which can lead to damage to human health due to the incorrect use of chemicals.

Three techniques are controlled by law on chemical pollution. Firstly, the use of chemical products that are hazardous to the environment is directly prohibited or restricted. Use, import, distribution or sale of such harmful compounds may be criminal, while less risky chemicals may be subject to licensing and permission procedures. This type of regulation is generally not designed only for aquaculture, but is contained in fundamental environmental laws or in more particular acts concerning the use of chemicals. Legislation in certain countries like as China, Thailand and Vietnam has been approved, including the restrictions on the manner in which these chemicals are utilized in aquaculture (Pandey et al., 2014).

A licensing system for wastewater disposal is developed as the second technique to regulate contamination with chemicals. Under the general environmental or water legislation, licenses are most likely to be regulated.

A third technique of regulating the use of chemical substances is to prohibit the direct use of certain medicines, chemicals and hormones which can adversely influence fisheries' physiological functions and welfare. Drugs, chemicals and hormones are sometimes seen as compensation for malfunctions or general lack of welfare in aquaculture enterprises. In Norway, the legislation dealt with the problem by prohibiting the use of chemicals to substitute for malfunctioning water systems.

**Figure 7.10**: Reducing the risk of aquatic animal disease outbreaks and improving environmental management of coastal aquaculture in Vietnam.

*Source: https://enaca.org/?id=568*

The Responsible Fisheries Code of Conduct also recognizes the problem of chemical abuse in aquaculture. Article 9.4.4 stipulates that effective farming and fish health control techniques that favor hygiene measures and immunizations should be promoted by States. 23 States must also assure safe, effective and limited usage, and regulates the use of chemical inputs into aquaculture which are detrimental to human health and the environment, of therapeutics: hormones, medicines, antibiotics and other disease-control agents. Virtually every country acknowledges the potential risks of chemical abuse in aquaculture. In most circumstances, however, laws are not only worded in relation to aquaculture activities.

## 7.10 WASTE AND WASTE MANAGEMENT

The aquaculture sector can produce waste in numerous ways, like in other food industries. This waste might damage the processing environment and should be minimized to the maximum extent practicable. Countries should have disposal and storage systems, and these operations should ideally be regulated by law. Many countries have established national waste disposal regulations.

Chile, for example, mandates aquaculture operations to take action in order to prevent discharge in the environment of solid or liquid wastes and trash (Panigrahi& Azad, 2007).

In order to reduce the disposal of dead animals, the waste exhaust, and water pollution in and around the aquaculture plant, the Zambia Fish Act permits the Government to grant requirements of an aquaculture license. A waste that could have dangerous and damaging impacts on human health and the environment is prohibited by the Namibia Aquaculture Act. It also stipulates, in line with the criteria and requirements of the license for aquaculture, which waste shall be disposed of.

## 7.11 COASTAL AQUACULTURE AS PART OF INTEGRATED COASTAL MANAGEMENT

In several nations, the process of integrating the aquaculture industry into coastal area management is progressively gaining popularity. For instance, the 199828 Belize Coastal Zone Management Act notably covers suggestions for aquaculture that are to be addressed in Coastal Zone Management Plans.

The Act also creates a consultative board with technical specialists with an aquaculture background in its membership Likewise, Chapter 1 of the

Philippines' Fisheries Code of 1998 states that state fisheries and aquatic resources policy "will be supported appropriately by research, technical services and State guiding in the manner consistent with the concept of integrated shoreline management in specific areas of natural fisheries management."

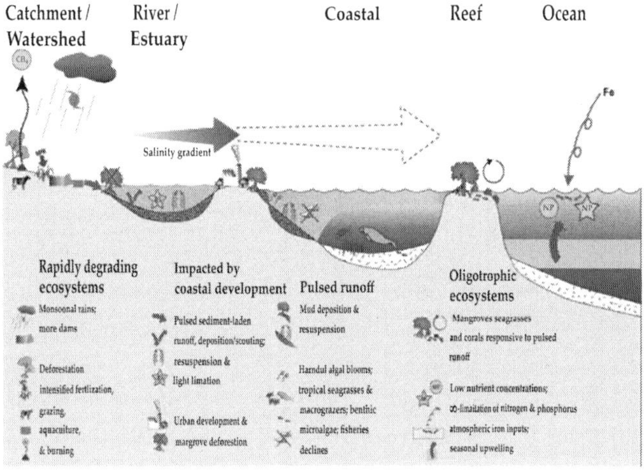


**Figure 7.11**: Integrated Coastal Zone Management.

*Source: http://cknet-ina.org/activities/read/integrated-coastal-zone-management/153*

Many tools are available for integrating aquaculture into coastal environments, for example, the application of EIA, the establishment of protected areas or exclusion zones, private ownership limits and indigenous acknowledgment of rights. Zoning, with land and sea regions being set aside for various forms of aquaculture, is one of the most essential instruments. The Chilean Aquaculture and Fisheries Law specifies special coastal areas for aquaculture. Aquaculture is prohibited outside of these locations.

The Tasmanian Marine Planning Act of 1995, which lays down the drafting of Marine Farming Plans in places where marine agriculture could take place, is another type of zoning. The plans are prepared with the public in consultation and take into consideration the physical appropriateness of future aquacultures, the present legal position and effect reduction for other users of coastal areas.

## 7.12 ESCAPEMENT

The escape of fish from a fish farm, known as escapement, has become ever more vital to monitor. The rationale is that extinct fish tend to contaminate the environment genetically and cause species genetic degeneration. Regulations concerning veterinary and ecological safety here, such as the quarantine and use of established or foreign aquaculture species, are particularly significant. These guidelines can decrease the potential influence of introduced or alien species on the farm species, wild endemic species, and biodiversity of natural water bodies by being included in the aquaculture regulatory framework.

**Figure 7.12**: Alaska exceeds Canadian chinook escapement goal, decline remains a mystery.

*Source: https://www.ktoo.org/2014/08/12/alaska-exceeds-canadian-chinook-escapement-goal-decline-remains-mystery/*

Possible veterinary control and environmental safety rules include procedures to decrease the possibility of exhaust or, when exhaust happens, to minimize harm. This covers the reporting and notification of exhaust effects provisions. For instance, Norway, Canada and Zambia are countries that enact comparable legislation. When fish escape, the following issue emerges: who is this fish's owner? The owner should have the right to get fish from an aquaculture farm. This is also vital in order to avoid the exhaust system and to take preventive action against the proprietor of the fish that have fled.

In the case of escapement, Namibia Aquaculture Act covers ownership (Ottinger et al., 2016). The legislation only allows the licensees to own certain aquaculture products, as long as they are inside the limits of the site. When the products are released or escape from them, they retain the licensee's sole property, provided the licensee proves its identity. The Zambia Fisheries Law allows secondary laws to contain ownership rules.

## 7.13 REGULATIONS OF TRANS-BOUNDARY AQUACULTURE ECOSYSTEMS

The control of cross-border aquatic ecosystems is also connected to escapement. Often governed by international or regional legislation, these include collaboration in marine aquaculture with riparian countries. Aquaculture producers should have an obligation to prevent environmental concerns such as exhaust, wastes, illnesses and effluent from cross-border aquacultural habitats. It depends on the cooperation and cooperation of countries. CCRF Article 9.2 deals with the topic of protecting cross-border ecosystems of aquaculture. Zambia, Uganda and Ukraine are examples of countries regulating this issue.

## 7.14 TRADE AND HEALTH ISSUES

In relation to the development and conduct of aquatic activities, national aquaculture law should attempt to apply international norms in commerce and quality assurance. If a country considers the export to European Union (EU) markets of fish products from aquaculture products, a regulatory framework that complies with EU market criteria should be developed. If international trade is considered, the WTO standards must be complied with, including obligations under the World Trade Organization (WTO) health and phyto-sanitary agreements.

Though voluntary, the Code of Conduct for Responsible Fisheries also provides certain accepted international law standards, and it should be aimed at incorporating it into national law by the states which have approved the Code. Aquaculture goods are a vital source of revenue for many underdeveloped countries. The Code contains many dimensions of international trade, helping States to meet and obtain legally binding export criteria. For example, the Code establishes basic safety and quality requirements across the industry. Like the Code, the European Hazard Analysis and Critical Control Point Directive (HACCP) defines European

Union export standards. Internationally adopted regulations exist like the Codex Code of Practice for Fish and Fishery Products.

## 7.15 INTRODUCTION AND MOVEMENT OF SPECIES

The incorporation of exotic species into areas where they contend with or displace native species is one of the potential negative environmental repercussions of aquaculture. The laws and regulations should therefore include provisions on import, quarantine and distribution of non-indigenous species. These regulations must be based on trustworthy information and should expressly take into account the ambiguities and negative impacts related to new species introduction. A number of existing codes of conduct and regional guidelines provide sample approaches to regulating the introduction, exchange and control of the spreading of species, with a special focus on disease spread.

The spread of species can have a wide range of effects on the environment, the environment and diseases. First, aquaculture inventories from the wild for production can have adverse effects on native stocks and ecosystems in farmers. The introduction of non-native species that might escape into the wild could disrupt local ecosystems by bad genes or by invasive competitive habitat effects on existing stocks. The movement of stocks of species between farms also risks transmitting illnesses which can have disastrous effects on individual farms and on the entire aquaculture sector (Opiyo et al., 2018).

Initially, species movement licensing regimes will prevent unacceptably hazardous movements and allow individual evaluation before authorization is granted. By making certain categories of movement unlawful, these regimes will prevent adverse ecological impacts and the spread of disease. Another benefit of licensing regimes is that they can impose conditions requiring all significant movements to be recorded and facilitate the communication of that information to the relevant authorities. Monitoring movements allows recently moved diseased stock to be traced to other potentially infected locations.

## 7.16 FOOD SAFETY AND HEALTH

The protection of consumer health and safety must be governed by public health legislation for aquaculture goods, like other food products. The Responsible Fisheries Code of Conduct covers public health and encourages States to ensure the safety of food products in aquaculture and support efforts to maintain product quality and value in advance and throughout harvest, on-site processing, storage, and transportation. The Code also encourages States to take adequate measures to safeguard consumer rights to safe, healthy and pure fish and fisheries products. The Code urges States to set up and maintain adequate procedures for the protection of national security and quality assurance and to consumer health and prevent economic fraud.

Likewise, in many countries that import aquaculture goods, HACCP has been increasingly incorporated into the regulations on aquaculture. These legislative advances are at the forefront of the United States and many EU states. The use and enforcement of these systems at the aquaculture agricultural level is relatively novel whereas exporters may experience a progression in the installation of safety guarantees systems in the fish processing sector.

Since January 2002, the EU has mandated the marking of its place of origin and species on all fishing and aquaculture commodities for retail sale to end customers. Sure "ecological labeling," however controversial may give a market-based incentive for ecologically friendly production. In this respect, the Code of Conduct claims that ' States should guarantee that, through improved identification of the provenance of fish and fishery products treated, international or domestic fish and fishery product trades (including aquaculture products) shall be compatible with sound conservation and management practices.'

## 7.17 MITIGATING INVESTMENT RISKS

The legal framework has to 'conciliate the needs of the environment and facilitate an efficient, shared interest aquaculture business.' Therefore, environmental rules are essential to excellent aquaculture legislation, but if its execution imposes an overload of expenses on the operator, aquaculture operations will, in the short and long term, not be financially or viable. This hazard can be eliminated by legislation that ensures, considering external and societal costs, that the overall cost of the aquaculture operation will not go beyond the benefits it delivers.

The legal framework should ensure that environmental objectives are accomplished, without putting additional expenses on applicants, in order to promote the trade development of the aquaculture business. To do so, governments may have adopted a single screen strategy for the various permissions required for an aquaculture activity and the preliminary screening of applications in order to ensure that the environmental impact assessment is complete only for those with considerable environmental risk. Countries could also consider establishing a single body to encourage aquaculture and track applications in other government ministries.

## 7.18 AQUACULTURE AND ZONING

Greater awareness about the ecological consequences of aquaculture has prompted governments to allocate diverse uses, including public land for the purpose of aquaculture. Zoning is a good strategy for integrating aquaculture in coastal regions as already mentioned. Zoning governs the environment at the farm level, equilibrates competing interests and resolves conflicts in relation to coastal resource preservation, usage and exploitation. For example, in Hong Kong SAR, marine fish cultivation areas are specified and all marine fish cultivation activities must be licensed in fish cultivation areas of their respective origin (Otta& Hoque, 2020).

In Ecuador, local zoning agreements between shrimp farmers and local residents allow shrimp farming to continue alongside traditional uses. Likewise in Zambia, conservation-planning areas that specify the layout of land for fish farms have been set up. One way to improve the legal and regulatory process for offshore aquaculture is to incorporate the creation of specific sites for aquaculture leases such as a Marine Aquaculture Zone (MAZ). Marine zoning faces significant challenges that its land-based counterpart does not, such as boundary disputes, enforcement difficulties, and more frequent user conflicts.

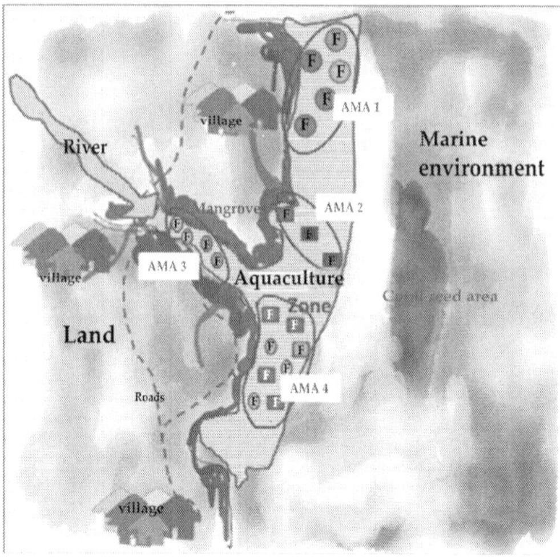


**Figure 7.13**: Schematic figure of a designated aquaculture zone (hatched area in blue color) representing an estuary and the adjacent coastal marine area. Individual farms/sites (F) owned by different farmers, are presented in different colors. Four clusters of farms illustrate examples of aquaculture management areas (AMAs), grouped according to a set of criteria that include risks and opportunities.

*Source: http://www.fao.org/3/i5004e/i5004e.pdf*

Zoning was a beneficial land-based strategy to identify locations suitable for the development of industries. The development of an aquaculture marine zone (MAZ) calls for the appointment of a national agency for the area, the issuing of leases to the seabed and water column, and the issuance of licenses that reflect the concerns of other relevant national authorities. Since the number of suitable locations is possibly limited, zoning of aquaculture promotes accessibility of natural resources and enhancing the possibility that aquaculture is successful.

The zoning also facilitates the weight of site labeling with area buoys and individual cage lights, which reduces theft, vandalism and liability problems related to boat cage collisions. Zoning also reduces the possibility of reef fish being attracted by the cages being caught by recreational fishers. Future aquaculture producers have founded a privileged commercial zone in Madagascar, known by the name of the "Régime de Zone Franche," business and monetary incentives. In practice, though, aquaculture areas have been recognized since the beginning of the 1990s, and potential entrepreneurs are encouraged to develop aquaculture activities in these areas by gaining the 'preferential trade status' that enables public access to financial and economic incentives.

## 7.19 SIMPLIFICATION OF ADMINISTRATIVE PROCESSES: SINGLE WINDOW APPROACH

The use of public authorizations for the control of aquaculture enterprises is an important characteristic of aquaculture legislation. Authorizations are subject to particular criteria, and can be made by licence, concession, or lease. They provide a suitable foundation for governments to regulate only limited natural resources and allow governments to include aquaculture farms into their coastal integration plans. The authorization procedure may be complex by the aquaculturalist, as a number of laws and agencies are involved in the licensing and management of an aquaculture project.

This is especially the case if it involves public access to land and water. The fisheries and aquaculture authorities are not just involved; they are also involved with land-use planning agencies, water bodies, health agencies and environmental authorities. A number of documents, such as land concessions, water licenses, effluent discharges and other sorts of environmental licenses, are typically required before an applicant can develop or manage an aquaculture farm (Ngajilo & Jeebhay, 2019).

In the field of aquaculture law, the issue now lies in removing legal and administrative hindrances and improving the profitability of aquaculture operations. The trend is toward the unification of licensing requirements and the streamlining of approval procedures through the creation of a single or lead aquaculture agency that controls the application process. The initiatives that have been advanced to streamline application procedures are promising. For example, in Chile, only one application for aquaculture activities is required under the Aquaculture Concessions and Authorisations Regulations. In Mexico, a special office for aquaculture development handles all the required permits, concessions and authorizations for establishing an aquaculture farm.

In Madagascar, the Prime Minister's Office created a Special Administrative Unit, and it was passed to the Ministry of Economics and Social Recovery as a one-stop investment shop.

(a) joining forces with technical departments, informing potential investors about the technical, administrative, legal and social environments as well as investing opportunities across the country,

(b) helping investors in the implementation of the investment rules and regulations and processes and

(c) receiving funding from the Office in developing and implementing the commercial aquaculture activities and also gain the status of free trade enterprises.

Malawi has taken a stand that promotes unrestricted investment in aquaculture ventures. The Malawi Investment Promotion Agency publications highlight aquaculture investment. This transformation places aquaculture on the same business level as agriculture. An informal single window technique is used by the ministerial committee to approve foreign investment proposals comparable to those outlined above.

## 7.20 FINANCIAL INCENTIVES

Aquacultural regulations reflect the awareness of the obligation to provide a person with security, irrespective of size, where public land is concerned. To this purpose, farm statuses for agricultural use of public land, prolonged or renewable lease periods and private-sector involvement in maritime tenure systems have been established in several nations. In Vietnam, guidelines have been laid down to grant 'stable and long term use' for aquaculture to households and people. The rental is renewable and extends for up to twenty years. In several emerging countries in Asia and Latin America, long-term leases and concessions (up to 10, 20 and 50 years) are also being created so as to:

- establishing a persistent contact with manufacturers and supervisors over time
- generalize the notion of sustainable development;
- guaranteeing the aquaculture industry holding such concessions the future and permanent supplies.

The right of the government to revocate such leases/concessions is limited, and can only be employed when management plans and/or the relevant laws are seriously violated. Wide lease termination privileges are likely to disappear for poorly stated causes. This growing trend, though, remains equivocal. In practice one should look whether or not, despite extra security being offered:

- Such leases/concessions tend to disregard the monopoly attitude towards ownership of aquaculture;
- Distorted competition;
- Other people have limited access to aquaculture. The Mozambique Investment Law 1993 provides for aquaculture incentives in which investment in aquaculture is given priority to government incentives.

Mozambique's investment law does not include provisions restricting domestic or foreign commercial or public aquaculture investment. Except where provided for by the Council of Ministers, Articles 11 and 12 of the Law stipulate that all economic activities are open to the private sector. A Promotion Center for Investment was established in 1982 to help the Ministry of Planning coordinate, evaluate and support national and foreign investment. An application must be submitted to Investment Promotion Centre, with the aim of achieving a tax exemption and benefiting from incentives. Tax incentives may include the exclusion or reduction, for example, the industry contribution, of import, movement and consumption taxes. Due to the great number of entities involved in gaining the essential approvals to conduct aquaculture operations in Mozambique, licenses and tax procedures have been proposed to be somewhat complex (Narbonne et al., 2021).

Malaysia, the Promoting the Investment Act 1986 permits to publish a list of activities or goods that can be promoted by the Minister of International Trade and Industry, so improving aquaculture development. In cooperation with the Minister of Finance, a company wanting to engage in promoted activity might be awarded a pioneer status certificate by the Minister. Companies having the status of pioneers shall have the right to a five-year investment tax allowance since the commencement of production, exempting them from income tax for a five- year period of 70% of their income.

## 7.21 SOCIAL IMPACTS

The apparent social impacts of fish farming are as perplexing as their environmental impact. Social concerns can be overriding grounds for disclaiming or regulating fish farms. Access to resources has traditionally been accessible to all users in many countries. Private fish farms are therefore considered to be inconsistent with traditional open access. Fish farming is thought to reward investors and fish farmers disproportionately.

In addition, the economic benefits of fish farming are restricted and do not stretch to local people because fish are grown for exports. Important socio-economic repercussions are conversion, expropriation, the privatization of mangroves and other lands, water and soil salinization, diminished food security, coastal population marginalization, joblessness, urban migration and social conflict. These repercussions often follow the development in many countries of aquaculture activity. Conflicts frequently arise when aquaculture farms are cutting off traditional methods of access to coastal resources by community people. This means that, if farms are huge, towns no longer have easy access to resources that they once used. Obtaining land and water access is typically a vital aspect in rural communities' survival and livelihood strategy. Social disagreements over rights to property frequently lead to a lack of definition of rights.

All land tenure difficulties are resolved as a serious problem in order for the aquaculture business. Many rural communities resisted water transfers to foreigners, claiming that such transfers may impose severe economic and social costs on the community. Unqualified aquaculture workers are usually from the zone where the farm is located, and employment conditions can be restricted. Because these people have few other jobs, there is limited competition in the labor market for their skills.

## 7.22 CONCLUSION

Aquaculture was seldom addressed in any implementation provision until the 1990s. If aquaculture is covered by a basic fisheries law, law enforcement sections have typically just been drawn out to take fisheries into account. Improvement in recent fishing and aquaculture laws was clearly made. All aquaculture practices without a license, the discharge without authorization of wastewater from aquaculture facilities in many countries lead to some kind of stipulated punishment. Generally speaking, the enforcement systems employed reflect conventional penalties systems, including administrative measures and fines.

Moreover, the right to visit, inspect and search aquaculture operations often appears describing the rights, duties and obligations of enforcement agencies. Nevertheless, it remains difficult to execute enforcement methods due to the duplication of legislation and organizations involved in aquaculture. Limited finances and insufficient personnel, which mean that often aquaculture rules are not fully enforceable, are a big challenge, particularly in poor nations. Consequently, alternatives are being examined other than the conventional 'command and control' systems in order to promote farmers to use resources more efficiently and be responsible for mitigating or limiting environmental damage produced by their aquaculture operation (Muhala et al., 2021).

The employment of economic mechanisms such as subsidies and tradeable licenses looks particularly to be of considerable importance. This evolution is illustrated by the 1998 Philippines Fisheries Code (PFC). The Code applies the responsible authority to "encourage environmental compliance and promote sustainable management practices" by designing incentives and disincentives, such as effluent costs, user fees and negotiable authorizations.

Chapter 8

# GLOBAL TRENDS IN INTENSIVE AQUACULTURE

## CONTENTS


8.1 Introduction .......... 192

8.2 Factors for Aquaculture Development .......... 193

8.3 Production Relating to Inputted Energy .......... 196

8.4 Tackling the Obstacles to the Development of Aquaculture .......... 198

## 8.1 INTRODUCTION

In the past 50 years, global aquaculture, with a value of $98.5 billion (including water-plants) in 2008 and a figure of around 50% of the world's fish food supply, has increased tremendously, to over $52.5 million in 2008. Asia is in the lead with 89% by volume and 79% by value, with Chinese being far the greatest producer (32.7 million tonnes in 2008). A multitude of variables, including current aquaculture practice, population and economic expansion, a looser regulatory environment and increased export options, have pushed significant growth in this region. The expansion of aquaculture in Europe and North America was rapid in the 80s–90s, but after that period it had slowed, perhaps because of legal constraints on sites and other competing considerations. Aquaculture growth rates were 6.9 percent a year between 1970 and 2006 but seem to be slow (average 5.8 percent between 2004 and 2008). It shows the typical trend of acceptance, followed by rapid growth that can be seen at the national level, and that is then slowed down by rising competition.

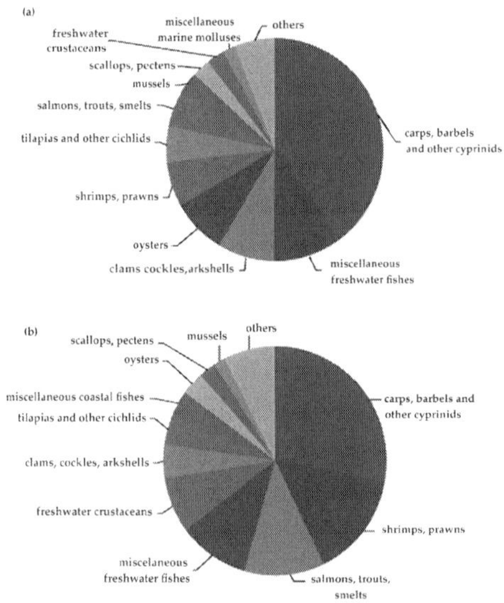


**Figure 8.1**: In 2007, a limited number of substantial countries, like Thailand, Spain, and Canada, underwent contraction of production. This was mostly due to market and competitiveness; however, sickness and unique natural events could also take part in one-year results. Overall, these losses represented 1.6% of global supplies (i.e., more than compensated by growth elsewhere).

*Source:* https://royalsocietypublishing.org/doi/10.1098/rstb.2010.0170

Relatively low output countries, such as Lesotho (6550 percent), Rwanda (909.6 percent) and Ukraine (580.8 percent) have the largest relative growth rates between 2006 and 2007. While this may serve as a useful indicator of new initiatives, a lesser percentage increase has a stronger influence in countries with already significant output. For example, China's growth of 5.2% represents 52.3% of the global supply of aquaculture for 2007. Vietnam, which accounted for 16.7% of increased aquaculture production at a growth rate of 30.1%, was the second most significant country in this regard (Mohapatra et al., 2013).

## 8.2 FACTORS FOR AQUACULTURE DEVELOPMENT

With respect to these factors, it is informative to investigate individual aquaculture industries. Market request (and competition), environmental resource availability, developing or transferring adequate technology as well as a favorable business climate that enables companies to profit from their industry investments are the main considerations. There are however numerous cases of a development failure, in Africa and certain areas of Latin America, because of the lack of incapacity of well-developed markets to reach them due to problems of infrastructure, including lack of acceptable export quality control.

### *Participants in the sector*

The whole aquaculture market is quite diversified and fragmentary, with yearly sales of more than US$1 trillion from smallholders' ponds in Africa that supply few kilograms of fish annually to worldwide companies. In 2006, an estimated 9 million persons, approximately 94% in Asia, participated in fish farming. The average output per person a year was 5.96 metric tons, but this ranged from 0.57 to 161.22 metric tons in highly industrialized Norway, where aqua-cultural methods tend to be labor-intensive (Mishra et al., 2017).

To most, aquaculture is a more restricted spectrum of economic activity in a single coastal or rural area and is especially important as a source of subsistence and cash crops in countries like Vietnam, India, and Bangladesh. There are also a considerable number of small- and medium-sized companies and single traders in Europe, with 13 139 entities, averaging full-time and sales of approximately € 270 000. Nonetheless, some species, particularly commodities items, which are marketed internationally, have strong trends towards consolidation and industrialization. For example, four corporations now account for 70 percent and two for more than 50 percent of the output value of Scottish salmon.

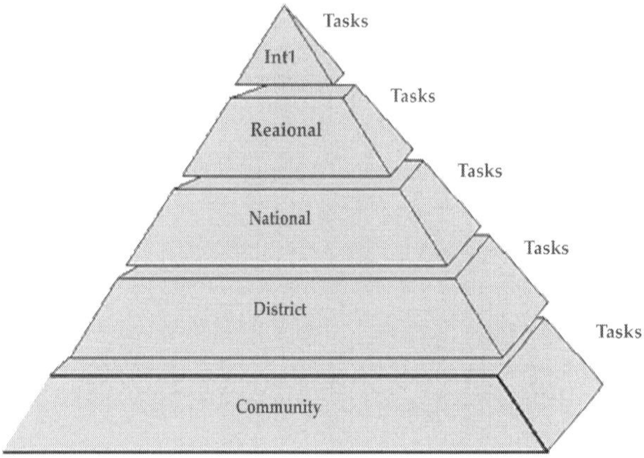


**Figure 8.2**: The structures of the market chain are critically linked to the viable size of the company. Europe's smaller businesses tend directly to market consumers and local restaurants and hotels, earning a considerable premium at normal wholesale market rates. It is not a solution that will saturate local markets for somewhat larger companies.

*Source:* http://www.fao.org/3/ab412e/ab412e32.htm

When producers are in competition in bigger marketplaces and there are minimum purchasing volumes imposed by much bigger purchasers, economies of scale become more essential. Producer cooperatives have been established periodically, where consolidation in locations is most often not physically practicable, that smaller enterprises can work with larger consolidated market chains.

Small-scale producer in Asia is similarly affected by international market chains. In Vietnam's catfish business, for example, consolidation is noticeable initially in processing, but is progressively being spread to production, pushed principally by the application of Western quality standards. In Asia, many small enterprises still have complicated chains. Efficiency is achieved by flexibility and service quality specialization and competition. A notable example is the production of live marine fish in China, Hong Kong, Korea, and elsewhere in the South-East Asia in restaurant and specialty markets, with values quite different from those of the Western marketplaces.

## *Interaction of resources and dependence*

Whereas the rise of aquaculture is profitable in many ways, strain is mounting on natural resources inputs, particularly water, fuel, and nutrition, while sites are also a problem in a broader sense. The issue also concerns the usage and impact of environmental services, especially in the dispersal and treatment of agricultural effluents. In terms of their need for these resources, aquaculture systems are highly different. Water use in freshwater systems is mainly a concern (the category trout and salmon cover a mix of both seawater and freshwater). These findings contradict those of Bosma and Verdegem (2009), who calculated total freshwater aquaculture water withdrawal at 16 900 $m^3$ $tonne^{-1}$ (excluding water returned to the aquifer).

A variety of systems is employed by freshwater agriculture, from static water ponds to high-flow vessels. Water consumption does not equate to water intake because most entail water intake from the environment and a post-production effluent outflow. However, the water quality is generally decreased, and evaporation and drainage may lead to water loss. Pond systems may lose 20 percent of their volume per day in tropical countries, in the worst situation. However, water management may also be assisted by pond aquaculture by catching and keeping surface water (rain and run-off) that could otherwise be lost from local agro-ecosystems or causing damaging floods (e.g., in the Czech Republic). Water reuse and recycling technologies can significantly cut use, although usually at the cost of increased energy inputs.

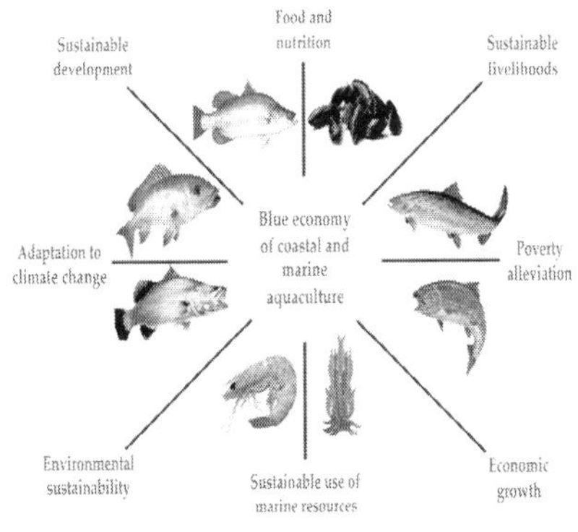


Figure 8.3: With demand on freshwater supplies currently increasing, a further expansion in aquaculture can be projected to use more plentiful brackish and marine waters. Environmental challenges are no less difficult, however.

Source: https://www.aquaculturealliance.org/advocate/the-blue-dimensions-of-aquaculture/

Most freshwater aquaculture uses semi-intensive techniques, employing a mix of natural and inorganic fertilizers and complementary feed, for their performance, based on regulated eutrophication. Additional fresh-water use (up to 3000 m3 tonne−1 according to Verdegem&Bosma, 2009) is made of feed ingredients for aquaculture, especially grain or similar crops. The solid wastes generated by such systems are frequently used in other crops as fertilizers. Solubilized nutrients can be often lost through required regimens of water substitution and can be a source of difficulty in places with a large production of aquaculture or oligotrophy or mesotrophic. Better optimization of water- and feed-management freshwater systems might quadruple productivity without increasing freshwater use (Verdegem & Bosma, 2009).

## 8.3 PRODUCTION RELATING TO INPUTTED ENERGY

More emphasis is given to the energy costs and the associated consequences of carbon emissions from aquaculture activities. Analysis should distinguish between direct energy consumption and a more comprehensive approach for auditing energy inputs, i.e., fuel and electricity spent directly in the process of production. For example, they can include industrial energy (energy utilized in equipment manufacturing and supply, feeds, and other energy inputs) or embedded energy, which also takes photosynthesis, sunshine, energy, or calorific values into account. Another issue is whether energy sources are renewable.

### *Pressures on the coastal zone and impacts on the ecosystem*

Aquaculture is regularly confronted by other users of this resource, most notably in coastal areas. However, other users, such as shrimp farms, which have been scrutinized for the overuse and destruction of mangrove resources and for other environmental repercussions and severe disease concerns, also have been affected by aquaculture. The wider worth of these settings as ecosystems are now recognized and adequately protected in most regions, but a lot has yet to be done in terms of reconstruction of destroyed areas.

More recent moves by the inland shrimp business also have generated difficulties in agricultural soils with the incursion of salinity (Balcázar& Ma, 2007).

Developing marine fish caging production has also given rise to concerns about broader consequences on the environment, ecosystems, and biodiversity. These are difficult to detect at modest developmental scales save for local alterations in sediments underneath the cages. The potential for larger-scale growth is that fertilizer or chemical wastes will be released directly to the environment or the effects on wild populations of the escaping fish or disease spread.

Another very immediate issue is typically a conflict of interest, like boating and sailing, recreation, seascape preservation and wildlife conservation. In Europe, these challenges are considered during or in increasingly developing coastal zone plans during the licensing process. The coastal pond and pump-shore tank systems face similar problems. Recirculated water systems can overcome several such restrictions but have been unable to compete financially except for more specific applications.

## *Feed*

Low protein, cereal-based supplementary meals and organic fertilizer are the main feed requirements in the bulk of freshwater fish production. Most crustacean, marine and other aquaculture intensive fish farming require a higher-quality diet that generally includes seafood and fish oil. Some aquacultures, especially tuna and marine cage crops in Asia, depend directly on wild pelagic fish with relatively low market prices.

This method turns fish protein for human consumption from low to high value. The effectiveness of this however is both an environmental and a social justice issue (e.g., consumers of farmed salmon and shrimp may effectively outcompete rural poor for this fish resource). In intensive animal husbandry, in particular pork and poultry, fish meals have always been used, thus aquaculture doesn't suffer. However, aquaculture takes up a rising percentage of this resource as replacements for cattle and poultry are more easily obtained.

**Figure 8.4**: Aquaculture will proceed to utilize most of the fish meal and oil produced unless alternatives for better values continue to evolve, but it will not be enough to fulfill increasing requirements for aqua food ingredients.

*Source:* https://www.bentoli.com/fishmeal-aquaculture-industry/

For at least the last 20 years, wild-caught supplies of fish meal and oil have variated at approximately 5-6 million tons a year, indicating that this is likely to continue. In 2008 however, around 90% of the world's fish oil and 71% of the fish meal were consumed in aquaculture activity. Omnivorous and herbivorous (carp and tilapia) feeds are often made with fishmeal or fish oil even if they are not needed for nutritional reasons. However, the rapid growth of carnivorous crops like cobia and catfish pangasius could raise the stress on the supply of fishmeal and oil. Improving the efficiency of food conversion with improved feeding technology is an overarching element that has substantially affected needs for fish meal and oil.

## 8.4 TACKLING THE OBSTACLES TO THE DEVELOPMENT OF AQUACULTURE

The emphasis is on enhancing resource use efficiency through implementation and coordination or more technology solutions through progress in engineering and bioscience. The wider economic and social issues, especially industrialization, urban and exchange rate pricing (energy) and consumers' demand, are crucial and influential for both methods.

## *Approaches of integration*

In the future, the aquacultural industry has an essential role to play in integrating aquaculture, fishing, agriculture, and other productive and ecosystem management activities. Ranching, agriculture/aquaculture, integrated multi-trophic aquaculture (IMTA), and connections with renewable energy projects are some of the techniques used. Integration is a crucial feature of the 'Aquaculture Ecosystem Approach,' which 'is a method for integrating activities in a broader ecosystem to encourage the sustainable growth, equity, and resilience of interconnected social and environment systems.'

Although capture fisheries and aquaculture are frequently thought of as different operations related simply by their market destinations, there are several significant system links between these two types of aquatic output. These include the interdependence for the provision of fish products to aquaculture feeds, the function of aquaculture resources in encouraging and improving fishing catches, as well as the development of controlled ecosystem approaches linking aquaculture and fisheries into one-space units.

Enhancing aquaculture fisheries comprises a variety of systems that integrate fishery and aquaculture characteristics. Improvements include the discharge in open waters of cultivated fish with the purpose of immediately improving fishery fishing or helping to restore depleted stocks (Barman et al., 2013). Examples include large-scale, cultivable fisheries in Asian reservoirs for big carps, ranching for Pacific salmon, upgrades to scallops in New Zealand and Japan, as well as many small-scale systems. Improvements may also include habitat and environmental changes to improve the productive capacity of wild or cultivated stocks and broaden private ownership over such assets. These improvements are also possible. For example, there are classic animal farming methods that recruit private coastal ponds or rice fields and modern developments such as 'free fish farms at sea' where fish are used to feeding stations.

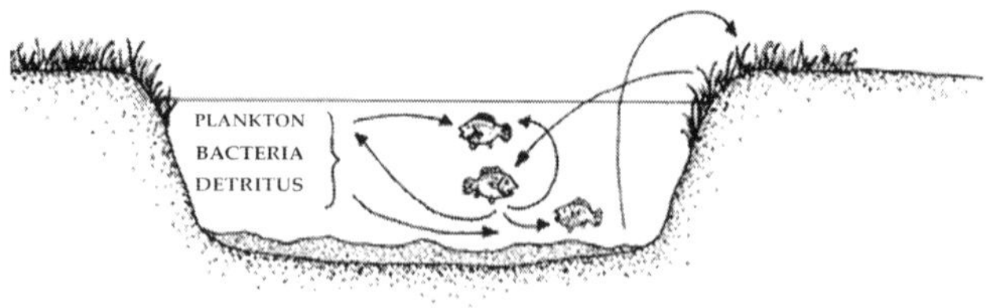


**Figure 8.5**: In the same bodies of water, or by some other water-based connection, integrated multi-trophic aquaculture systems may be described as cultivation systems using species from various trophic levels (for land-based systems). The scale must not be large if the species are cultivated and the quantities produced are compatible.

*Source:* http://www.fao.org/3/Y1187E/y1187e08.htm

Improvements in aquaculture fishing may offer significant ecological and genetic risks to wild stocks. These hazards can be decreased through productive improvement, but not totally avoided by ecologically and genetically separating ecological releases of wild and wild populations (for example, by release and habitat in environments that do not interact with wild fish) (e.g., by maintaining captive brood stock and releasing sterile fish). Where this is physically achievable (i.e., fishing is not un-selective for cultured/wild fish), selective harvesting of discharged cultured fish may further limit influences on wild stocks. Further influences on the wild stocks and natural ecosystems could be achieved by environmental alterations and feeding. In efforts to restore wild stocks, cultivated fish are to deal with wild fish, and stock and genetic management must be conducted with special care to prevent adverse repercussions on the wild stock depleted or even at risk. Supplementation and captive breeding programs may help the conservation and restoration of such stocks, but, in this situation, management tactics differ greatly from those used to improve output.

Water is the transmission vector of nutrients for dissolved and particulate wastes, in all situations, the discharge of one species as food for other species at a less tropical level. The acceptance of environmental costs (environmental economy) within the system could significantly alter these. New products of marine species could be developed. Other restrictions include existing regulations limiting the future expansion of aquaculture or the possibility for unforeseen system interactions. The potential of the technique is however obvious to address sustainability goals. Dey et al. (2010) assessed the influence on the Malawi smallholder IAA-supported program of WorldFish, and found that technology adopters achieved a rise of 11% in total factor productivity (TFPs), a rise in technical efficiency rates of 35%, a 134% increase in farm income per hectare and a 60% increase in overall income as compared with non-adopters Non-adopters had larger wages from off-farm activities, but overall family labor returns were higher and household incomes higher (almost 1.5 times higher). This suggests that IAA could contribute to poverty reduction and improvements in livelihoods in Malawi and presumably in other similar nations (Michael et al., 2014).

Further issues have been identified where IAA is practiced on a wider scale with commercial products. Quality can for example vary with contamination problems, such as pesticides used with irrigation water or off-flavor taint that limit acceptance and certification, particularly for overseas markets. IAA's advantages have primarily been centered on food production to yet, but more sustainable use of freshwater and energy could become similarly essential in the future. In developed countries, interest and activity are growing with tiny aquaponic systems, which mix freshwater aquaculture and hydroponic gardening, usually herbs and salad. The vegetable crop is mostly nourished through fish crops with nitrogen waste. These systems have not proved economically appealing because of scaling challenges, but they are suitable to produce food 'back-yard,' which is a strategic candidate to increase the production of sustainable food.

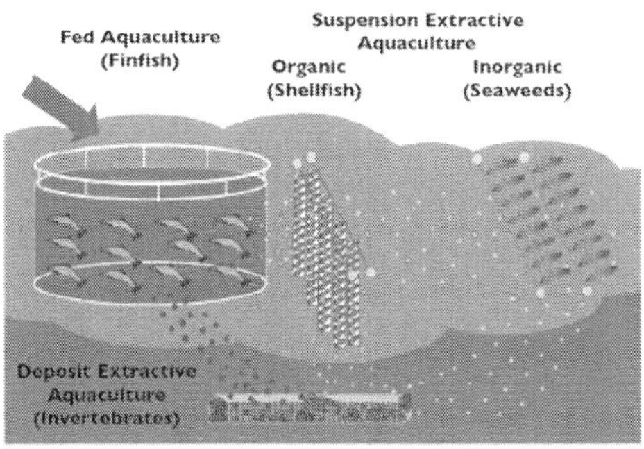


**Figure 8.6**: In poor countries, integrated aquaculture/agriculture provides a mechanism to broaden and increase productivity for rural systems. The method of culture differs from the single-culture approach, which typically poses too high a risk for poor farmers. Integrated systems profit from synergy between the different components and have production variety which gives rise to relative environmental health.

*Source:* https://www.aquaculturealliance.org/advocate/integrated-multi-trophic-aquaculture-part-1/

## *Solutions to issues with resources*

Proteins and other nutrients generated from fishmeal (essential amino acids) replacements are nutritionally straightforward and significant progress is being made in this area over the last 30 years. The problem for the supply of proteins is mainly the economy, formulation, and ongoing evaluation of potential new protein sources. Up to 75% of the fishmeal in normal feed can readily be replaced even with carnivorous species (high dietary protein levels, sensitive to feed palate).

Fish meals are superfluous for allergic and herbivore animals, and are only now employed, since they are economically viable. Whether the carnivore species are fed on 'vegetarian' diets is an ethical problem for fish welfare. Further, soya bean produces enteritis in Atlantic salmon, and it may have an impact on the welfare of fish in general for plant proteins (which contain large ranges of nutrients and non-nutrient fractions not usually exposed to fish).

Krill is another possible source of feed protein and oil (a collective name for a group of roughly 80 species of tiny, pelagic, shoaling crustaceans). Nutritional problems with the quality of the product (fast spoiling) and its fluorine content were successfully tackled and suitable capture and processing procedures were established. CCAMLR3 predicts that an estimated 1 teratonne of krill meal would be provided with an annual amount of 32,000,000 million tons of krill oil from Antarctic waters. Aquaculture, however, faces intense competitiveness for krill resources since high-quality krill has been increasingly used for direct human use and manufacturing of medicinal krill oils. There should also be consideration of the possible impact on marine food webs (Marques et al., 2020).

Aquaculture is still largely produced from wild or freshly domesticated inventory. The lack of genetic engineering and poor nursery techniques has greatly affected the performance of several agricultural species through genetic drift, inbreeding, and uneven hybridization, especially in poorer nations. Reducing performance and sustainability implies that hatchery inventories typically must be replaced routinely with wild fish or better-managed stocks from other farms. In contrast, well-managed selective breeding schemes, performance and quality have continuously improved.

In approximately six generations, Atlantic Salmon breeding businesses have shown more than 100% improved growth, with considerable increases in the resistance to disease and delays in the commencement of oocyte development. The great majority of farmed Atlantic salmon eggs and smolts are now produced by breeding businesses and similar procedures in certain other species are now deployed.

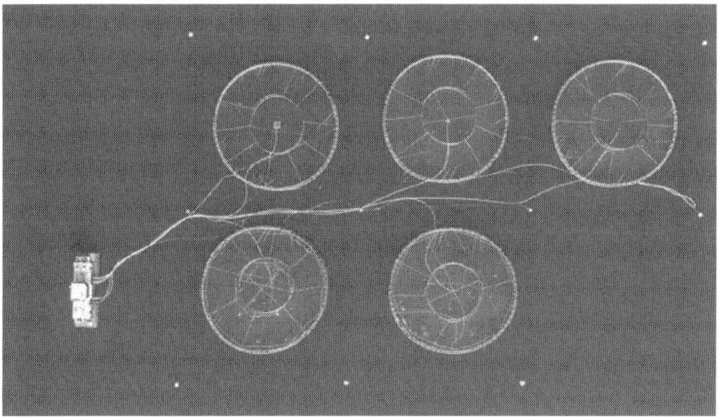

**Figure 8.7**: Selective reproduction can enhance the annual performance of farmed fish stocks for a variety of features, but it is frequently still essential for these fish to use several other strategies to realize their maximum potential. The end yield can be greatly decreased when sexual maturity in production fish can become more aggressive, stop growing, lose condition and become more disease sensitive. One or another sex is preferred in many animals, e.g., because in crop sized, it grows quicker or remains immature. Females generally develop later than males in salmonids.

*Source: https://www.innovationnewsnetwork.com/sustainable-aquaculture/580/*

All-female breeding is now almost routinely employed in Europe for rainbow trout raised to portion size (more than 300 g), as females are always juvenile at harvesting. All-male production is favored in tilapia because, while the males develop, the lack of females prevents the undesired generation of fry that occurs in mixed-sex on-growing setups. Sexual maturation is harmful to some species and under specific conditions. Sterile fish may be produced with chromosome set manipulation techniques, which produces animals with three chromosome sets, called triploidy induced. This can be avoided. This strategy is presently being employed in the cultivation of enormous rainbow trout (above 3 kg) which are still in primary condition and growing. Triploidy is also commonly employed in all-year oyster production.

Regarding health management and welfare, in some intensive aquaculture systems, severe disease has proven an important limit on efficient production. In recent years, significant gains have been achieved in understanding the etiology and epidemiology of fish diseases, and aquaculture producers have significantly improved their methods in numerous nations with greater emphasis on fish welfare. New drugs and vaccinations have managed to control many major infectious diseases, and this applies to bacterial diseases. However, new medical concerns are arising, and rare diseases are becoming more widespread beforehand, which requires ongoing alertness and solution research (Lulijwa et al., 2020).

For bacterial fish infections, vaccines were quite effective, when resources are available to create them, but success with viral diseases has been more restricted. However, the use of recombinant DNA technology, particularly for pancreatic necrosis infection followed by a DNA vaccine, was amongst the first cases to be treated, which seemed to be highly promising. Given that technique includes gene transfer, important safety and consumer acceptance issues need to be addressed. The use of proteomics and epitope mapping to identify the vaccine's antigens and to develop peptide vaccines is another promising approach. This strategy is perhaps appropriate for the prevention of parasites like salmon lice. Other techniques include the use of viral particles claimed to be employed against a grouper neurological necrosis virus or yeast-producing virus recombinant proteins.

Aquaculture uses and integrates technologies from other areas like fishing, water-treatment, or offshore oil in the building of cultivation systems. With the industry growing, however, more sophisticated technology including salmon industry well boats, is emerging. Recirculated aquaculture systems (RAS) and offshore cage technology are of great importance to reduce water resources pressure and minimize impacts on delicate freshwater or coastal environment.

**Figure 8.8**: Offshore cage aquaculture has no universally accepted definition. Sites are characterized in Norway by wave height, while offshore aquaculture activities are described in the United States as activities in the unique economic zone ranging from a three-mile coastal territory boundary to 200 miles offshore.

*Source: https://www.aquaculturenorthamerica.com/robotic-solutions-for-aquaculture/*

Generally, land-based RAS culture systems use confinement systems like tanks or fish races. Following waste removal and treatment, a portion of the water from the outflow is returned to the system. The degree of waste management and water recycling is mostly determined by the needs of the fish, environmental conditions, and available technology. The reuse of water allows farms a greater degree of environmental control, reduces water use and trash disposal and permits near-market production. RAS is primarily limited in its use to higher-value species or life stages due to relatively expensive capital expenditures, high energy dependency, and more complex technologies (especially hatcheries where control over environmental conditions is more critical and unit values higher). However, if economic conditions change, it may become a more competitive strategy.

Regarding offshore cages, near-shore systems encounter several issues, including aesthetic impacts, local environmental concerns, and space limits, which can be overcome by moving systems further offshore. Predation difficulties and illness risks might be significantly decreased in most circumstances. The offshore industry's development would allow for larger project scales and, as a result, increased productivity. Waste discharges would be more easily distributed, and interference with other interests like tourism and inshore fisheries might be lessened. Other issues and risks connected with intense cage-based aquaculture, on the other hand, would persist or even worsen.

Offshore farming is defined as being more than 2 kilometers from the coast, vulnerable to high oceanic swells, unpredictable winds, and limited physical accessibility, and needing primarily remote activities such as automated feeding and distance monitoring. For these purposes, offshore aquaculture systems must have sturdy facilities and related systems that are able to operate in a high-energy environment with minimal input. There are also significant problems with employee safety that boost costs over near-shore systems.

The vast scale and the quantity of new technology necessitate large-scale cage farms to be able to use them until farms and firms achieve a range of operations where offshore investment is feasible. Marine Harvest, the world's largest salmon farming corporation, has indicated its desire to register for and construct offshore sites, indicating that this is beginning to happen. This is to produce salmon, but numerous species have been advocated as possibly suited for offshore farms (based on biology, economy, and the use of cobia).

Chapter 9

# FACTORS AFFECTING INTENSIVE AQUACULTURE PRODUCTION

## CONTENTS

9.1 Introduction .......... 208
9.2 Physical Factors Influencing Intensive Aquaculture .......... 208
9.3 Markets .......... 216
9.4 Management .......... 218
9.5 Political Goodwill .......... 219
9.6 Society .......... 220

## 9.1 INTRODUCTION

The adoption of intensive aquaculture in the world has been, moving in an upward trend. Most individuals that are commercializing fish farming are finding it suitable to improve on intensive aquaculture in order to improve the size and the quantity of fish to be sold. This has been able to help in meeting some of the rising demands of fish all over the world. The high demands of fish are attributed to increased awareness of the nutritional value of fish meat. As a matter of fact, there are those that are currently replacing red meat with fish meat because of the various health implications red meat has on the human body. Just like any other aspect in the business world, intensive aquaculture is also seen to be influenced by a number of factors. There are those factors that have negative impacts while there are those that generate positive impacts. Some of the factors that influence intensive aquaculture include physical factors, socio-economic factors and technological factors.

**Figure 9.1**: The increased demand for fish all over the world needs that individuals start fish farming.

*Source: FOA of the UN.*

## 9.2 PHYSICAL FACTORS INFLUENCING INTENSIVE AQUACULTURE

There are many types of bacteria that may be found inside a fishpond or fish tank. Bacteria are microscopic and cannot be seen by the naked eye, and most of them range between 0.2 to 10 microns in length. Nevertheless, because of their huge numbers and their fast reproduction rate their biomass in a fishpond could be quite high. Most of the bacteria that are found inside a pond are spherical or rod-shaped and some of them could bc filamentous. These bacteria mostly occur freely but they are found in large numbers on the surfaces of organic matter that are on water. In most instances, bacteria obtain their energy by digestion of organic matter and for this to occur they require molecular oxygen for the process of respiration. These types of bacteria are commonly referred to as obligate aerobic bacteria because they require oxygen for them to survive, whereas the opposite types are referred to as obligate anaerobic bacteria which cannot live in the presence of oxygen and there are some that live in both circumstances and these are referred to as facultative bacteria (Luis et al., 2019).

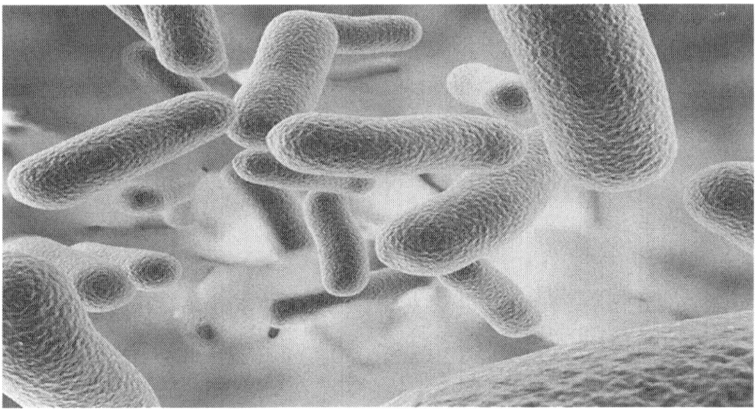

**Figure 9.2**: Bacteria are spheroid or rod-shaped; they are microscopic and cannot be seen by naked eyes.

*Source: POND Trade Magazine*

In the pond water as it enters into the pond it deposits some sediments into the pond and the pores found in the bottom soil of the pond are usually aerobic i.e. they contain dissolved oxygen. The bacteria that are found in these aerobic zones help in the decomposition of organic matter by oxidizing it to form other substances such as ammonia, water, carbon dioxide, other compounds and sometimes phosphates. This process is important in that it prevents the accumulation of products at the bottom of the pond and this will lead to a decrease in the volume of water. The oxygen will deplete at some point by at this instance the anaerobic bacteria carry on with the decomposition process.

In ponds, there is a process of nitrification and this is a process carried out by organisms from the genus *Nitrosomonas* that oxidize nitrite to nitrate and some from the genus *Nitrobacter.* With this process, it helps in the prevention of the accumulation of ammonia, which is toxic in the pond. As nitrification occurs, it leads to the acidity of the pond and this is why the farmer will end up treating the water with agricultural lime in order to obtain the required alkalinity for aquaculture.

As the bacteria will most of the time decompose the organic matter at the bottom of the pond some of the organic matter will take time to decompose and this will lead to its accumulation at the bottom of the pond over time. With all the accumulation, it will lead to formation of sufficient amounts of ammonia nitrogen, which will help in the growth of phytoplankton. With the growth of these planktons, it leads to reduced water volume and the fish's habitat will be drastically reduced. The phytoplankton will also reduce the amount of oxygen, which will lead to competition between the fish and phytoplankton.

**Figure 9.3**: Sludge formed at the bottom of a tank discolors water when disturbed and reduce water volume in a pond.

*Source: Joe Jaworski's- wordpress.com*

In intensive aquaculture there is the use of a lot of commercial feed, some of it will not be eaten, and these will lead to the formation of sludge. This sludge will include a mixture of uneaten feed, fish excretion and dead plants in the pond. The sludge will become a breeding ground for disease-causing bacteria. The sludge due to it being loose at the bottom of the pond will easily move when slightly disturbed. This causes the water in the pond to become discolored and disturb the fish sight. When disturbed it can easily lead to lowered visibility in the pond and there will be little or no light penetration into the pond and ultimately this will lead to fluctuation in the temperature of the pond (Liao et al., 2009). The sludge when left for long periods will increase in amount to impact the aquarium, especially for farmers practicing intensive aquaculture as there is a large number of fish in the pond and with sludge in the pond, it will lead to reduced volume of water. The low volume will make the oxygen in the water reduce drastically and it can lead to the death of some fish or in some instances, there will be stunted growth of fish.

The water quality delivered into a fishpond is also a factor in intensive aquaculture. Several factors need to be looked into when observing the water quality to be used in a fish aquarium this is because some fish require or survive at specific temperatures and for farmers with their own hatcheries the fish require specified type of water. These qualities to be observed include; dissolved oxygen, nutrients (nitrogen and phosphorus content), mineral content and pH. The temperature of the water being delivered is essential as the temperature can affect their feeding, behavior, growth and reproduction. Some scientists even believe that the temperature of water for fingerlings can influence them to change their gender. Dissolved gases are very important i.e. oxygen. With low dissolved oxygen, it will lead to the death of fish. Oxygen is important as it is involved in various processes such as respiration in the fish's body. Studies show that small fish consume more oxygen as compared to larger fish due to the metabolic rates being high in small fish. Coldwater fish have also been known to consume more oxygen as compared to warm water fish such as tilapia. This means the water being delivered to the pond requires to have adequate amounts of oxygen.

**Figure 9.4**: Discoloration of water due to algae.

*Source: Bunnycart.*

Algae always form on fish tanks or will form over time on fishponds. The algae will form when there is too much light and nutrients in the pond and this will make them grow rapidly. His is when the tank is getting direct sunlight and when there is too much fish food in the pond. This can also be caused by lack of water changes i.e. the water inside the pond is used for too long without being changed. When it is not corrected at the right time, it can kill the fish and in some instances, there will be an algal bloom at the top of the water.

Capital is a very important factor in intensive aquaculture. due to the fact that there is the use of commercial feed, large swarms of fish in ponds and use of machines in the business. Intensive aquaculture is mostly carried out by rich entrepreneurs and large companies or farms. The amount of money that one needs to set up the enterprise is quite high. The individual or company has to purchase land or pay hugely in order to hire the area in a coastline to set up the cages or the land to dig the ponds. For some farmers, they build huge warehouses and keep their tanks and/or aquariums inside. After establishing all these, they have to stock the fish inside the tanks, which is expensive because they have to purchase a large number of fingerlings. Some farmers set up their own hatchcries.

**Figure 9.5**: A lot of capital is used to set up an aquaculture farm.

*Source: BBC.*

There are machines needed for the whole process to continue smoothly. The farmer has to purchase water tanks for water storage, water purifiers for cleaning of water and removal of minerals and other salts, pumps so that water can be pumped to the tanks for storage and also to the ponds for use, oxygen pumps which will be needed to infuse oxygen into the water directly or they will pump the oxygen directly to the pond and/or aquarium for bubbling, fish processing area due to production of large numbers of fish they need to have areas to process the fish to be ready for the market, freezers for storage of fish when ready for the market.

Capital is very essential, as it will be used to employ the staff that will work on the farm. Individuals that are needed include technicians and mechanics so as to monitor and operate the various machines in the farm, aquaculture expert who is an individual who's an expert in the type of fish to be reared, the amount of feed to be supplied, the living conditions of the fish i.e. water quality and other factors needed for proper growth of fish, casual labor which will include people doing regular jobs to ensure smooth running and a manager to oversee the running of the farm and ensure that all activities required to be carried out are done on time and in the right manner. The amount of money an individual needs is large and it will determine the size of enterprise the individual or company will set up. Capital will be needed from start to finish i.e. maintenance of machines and wages of the workers, buying of feed and many others, which are very essential for the fish to survive (Li et al., 2011).

Technology is seen to be the major driver of many sectors in the world. The fish industry has not been left behind as individuals are using technology in improving the sector. The adoption of the new technologies that have been identified in the aquaculture sector has actually played a huge role in the development of the agricultural sector. It should however be noted that most aquaculture farmers do not only adopt new aquaculture technologies because they exist and are available to the public. This proves to be very hard news for aquaculture researchers especially when some of the proposed technologies developed by them are not adopted by the fish farmers. The decision for a farmer to adopt a particular technology is not an easy task, as the farmer needs to consider a number of aspects before embarking on the same journey. One major aspect is that most of the technologies prove to be very expensive for the aquaculture farmers to adopt which makes it quite hard for them to purchase.

In the last fifty years, there have been developments in systems and technologies that can be used in aquaculture. These developments have been very rapid that even the initial technology can only be seen as an invention too old to be used. These systems vary from simple facilities such as the use of family ponds where the fish are reared only for home consumption to high valued technology systems, which involve the intensive systems that rear fish enough for exportation. Most of the technologies that are used in the aquaculture system are more simple, which normally involves simple modifications done on the traditional ponds. Such modifications are normally done to improve the growth and survival of particular species of fish, which include improving the food and the food systems, improving the oxygen levels in the pond and protection of the fish from predators. It is not only in intensive systems that fish are protected from predators but all the systems require that the fish be protected from predators. There are cases of raising freshwater ponds for the purpose of raising herbivorous and filter-feeding fish are some of the technologies that are used all over the world. Individuals have been able to come up with a mechanism to allow for planting feeds for fish that are herbivorous but are being raised in the intensive system of fish production.

**Figure 9.6**: Fish farming technologies in intensive aquaculture are known to be the future of the fish agricultural systems to improve the quantity and quality of fish.

*Source: Aquaculture Magazine.*

Education has allowed individuals to understand the complex nature of the interactions of nutrients, bacteria and cultured organisms that are very useful in the fish development process. Through this, individuals have also been able to understand the interaction of the stated aspects with the advances that are in hydrodynamics that are applied in the pond and tank design. All these developments have allowed for the development of a closed system in the intensive fish industry. In fish farming, there are those fish that cannot be reared anywhere else other than in the seas and oceans. As such, the closed systems can be very helpful to keep them in one location. These systems have the capacity to allow for isolating the farmed aquaculture from the natural systems. This helps in minimizing the risk of genetic impacts, diseases and predators from the external systems (Labh & Shakya, 2014).

There have been developments in engineering that have been able to be used in the aquaculture sector that have been adopted from the oil rig construction systems. These have been able to increase the possibilities of progressive offshore expansion in the intensive aquaculture units. It is quite easy to recognize the economic benefits from the adoption of technology in intensive aquaculture. In recent studies, there have been cases of the development of fish feeds that will improve the aspect of feeding to make work easier and also reduce cases of water pollution. Traditional feeds are known to be always sinking and tend to contain low crude protein content. This is because; they are normally manufactured locally at the farms. On the other hand, improved feeds are known to be nutritionally balanced and are also known to be floating pellets.

## 9.3 MARKETS

In order to be successful in any business, marketing is a very important aspect. As a matter of fact, entrepreneurs are always advised on the importance of scouting for the market of their products before they can be able to embark on a particular business. Likewise, fish farmers need to be very proactive when they want to market their products either locally or internationally. The local markets are actually very important in the fish farming business given the fact that fish are products that are highly perishable. In the case where one has been able to locate a local market, it would be very wise for them to establish their system close to the market. Intensive aquaculture can be very expensive and thus the need to establish markets in order to be able to break even in the business.

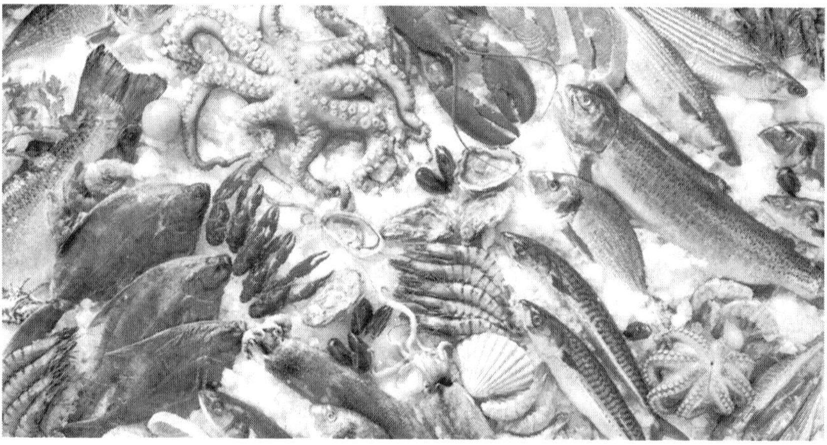

**Figure 9.7**: Intensive fish farms need to be established in areas close to the markets because of high cases of perishability and the need for freshness.

*Source: WGBH.*

In the north-central region, research was conducted and the conclusion was that there is always a perception the consumers of fish have. The belief is that, the farm-raised fish are fresh, are of higher quality and are healthier than the wild-caught fish. In the process of market development, it is always wise for the farmers to understand some of the important aspects such as careful handling, cleaning, processing, packaging, and transport and retail sales. These are some of the aspects that consumers consider before they decide to purchase the fish. With the increase in sales, the fish farmer can be able to increase the amount of fish reared and further increase their margins. In the case where farmers are able to take into consideration the above aspects in their fish production, they can then define their fish as a high-quality, high valued product. It is very important that as a fish farmer, one needs to look to develop niche markets where they can be able to sell even small quantities that are of high value. Genuine products are very important and the farmer needs to ensure his are even if it will not allow them to sell the products in bulk.

One needs to understand the individuals that are specifically going to buy the fish from the farmer. This may be either the direct consumer, restaurants or fish outlets. Another aspect is that the farmer needs to be sure that there is an actual market to supply the fish to. Another thing to consider is the pricing of the fish. This is the reason as to why there needs to be scouting before deciding where they sell. The price should be one that is favorable to both the consumers and the farmers, such that it is acceptable to the buyers and can benefit the farmer. The farmer should also sell at a price that allows him to have profits in order to allow for the expansion of the business (Kalbassi et al., 2013). The farmer needs to also know the risks that he may face in the business in terms of financial, biological, physical or social aspects. In addition to this, the farmer needs to establish a plan to sell his products. In a business, it is very important to set up a plan in order to allow for the proper establishment of the business and further know some of the issues he may face and how to go about the same. In addition to this, it is always important for the farmer to be ready for any uncertainties such as the fact that there are some cases where there is no market for fish and he needs to understand that there is no need to produce something that they cannot sell.

The input market is also an important aspect of fish marketing. Before starting aquaculture, in any system, it is very important for the farmer to understand how he will acquire the fish at the initial stage or when he requires replenishing the stock. The farmer should be able to spot the right person who has a better reputation in the business. When doing intensive fish farming, it is important that one grows the species that he needs without having cases of having other species that may be predators or those that may not be able to survive under the conditions that they have been put in given the fact that sometimes, the farmer adjusts the conditions to meet a particular species of fish. As such, the farmer needs to be able to ask around and gather information from other fish farmers in order to know the individuals that one can trust to buy the fingerlings from.

## 9.4 MANAGEMENT

In the fish farms, management is a very important aspect. It is not only in fish farming but also in any other business that management is an important aspect. A continuous proactive management process is something that is required in fish farms. In most cases, after a fish kill, most farmers ask the question "who would have thought?" O be a very proper manager in the fish farming process, it is important to be able to anticipate and be able to prevent some of the problems that are in the fish business. Intensive aquaculture normally involves the loss of the fish because of the loss of oxygen. There are also cases of diseases, predators, flooding, vandalism and even cases of thefts. When the manager of the business comes up with a proper plan and provides proactive management, the farmer can be able to avoid a lot of problems.

As a matter of fact, the owners of the business need to have knowledge of how to manage the fish business. Information is important in any business and intensive fish farming requires that the owners of the business have knowledge of what they are doing. It is also important to welcome new ideas in the business by attending seminars and having extension workers on the farms. Research has shown that there are extension workers who observed a number of aspects once they were able to work with the fish farmers. Consequently, they have been able to mention some aspects that are crucial to ensure the success of intensive aquaculture. The most crucial aspect that they observed was human resource. The claims of the extension workers were that personal commitment to the business and human resource were important factors that ensure the success of the intensive aquaculture systems. It is very important to recognize that hard work and the drive to succeed are some of the aspects that have driven intensive fish farmers to success. Some of the things that are necessary for the success of the business is knowing how to and aspirations to learn more on how to perform certain aspect. It is also very essential for the owner to know how to react to problems and be able to prepare for the same.

**Figure 9.8**: An intensive fish farmer needs to be hardworking and committed to the business in order to ensure the success of the business.

*Source: USAID.*

All these have been able to show that aquaculture requires hard work and commitment in order to ensure the success of the business. Another aspect is that it is important for the farmer to understand that fish are live animals and need that they are treated as such. To reduce the risk that one may face in the business, it is advised that one starts small. In addition to this, one needs to grow high-quality high valued crops in order to allow for a better market experience. Experience in business and having knowledge on the same are also very crucial aspects in intensive business production. Marketing is the big fish to fry as his is the point where money is made. It should however be understood that it always takes longer to make profits in the aquaculture business. This is given the fact that the business requires a number of things and that it is a high-risk business. With intensive aquaculture, there are times where the sales will be good but there are also other times when the sale will be bad. The fish may be attacked by a disease, which may lead to the farmer going through immense losses (Jana et al., 2018).

## 9.5 POLITICAL GOODWILL

This is one aspect that may be easily ignored but is one aspect that easily influences intensive aquaculture farming. As an intensive fish farmer, one would always want the government and the society to be behind their back. The government plays a very important role especially in allowing for the practice of fish farming using a certain system and setting policies to help the farmers in their daily business. With the government, markets can be able to be enhanced by the government policies such as the imposition of tariffs to encourage exportation of fish products and discourage importation of the same. With the government policies, the farmer can be able to know the types of fish that are allowed in the country to be reared. There are those countries that have banned the farming of certain fish species and thus it is important to understand all the governmental guidelines.

## 9.6 SOCIETY

Society includes the individuals that are within the vicinity of the business. These individuals can influence the business. As seen, farming fish in intensive system can be very expensive and as such, one needs to create a good rapport with society, as they are the consumers of the same business. The most important aspect is trust especially when one needs to see a boom in their business. As such, as an intensive fish farmer, one should be able to relate well with the individuals in the society so that the individuals in the society can be able to spread good things about the person. The society sometimes can prove very risky in the business, especially with cases of thefts and the cases of waste disposal to a water source. This can be very disadvantageous, especially if the farmer is dependent on the water source in his fish farming.

All the above-stated aspects that influence intensive fish farming as important especially when one is either establishing the business, continuing with the business or has an idea of expanding the same. To be a good fish farmer, it would be wise for one to consider all the factors mentioned above among others.

Chapter 10

# AQUACULTURE PRODUCTS AND MARKETING

## CONTENTS

10.1 Introduction.......................................................................................222
10.2 Aquaculture Products.............................................................................225
10.3 Crustaceans ..........................................................................................225
10.4 Finfish...................................................................................................230
10.5 Aquaculture Marketing .........................................................................236
10.6 Benefits of Aquaculture Marketing Products..........................................238
10.7 Features of Marketing of Aquaculture Products ....................................238
10.8 Marketing Channels..............................................................................239

## 10.1 INTRODUCTION

The aquaculture sector is among the sectors rapidly growing in the current day and age. This is as a result of the growing demand for aquaculture products with the increase in population. This rapid growth has resulted in significant changes being made to the perception and marketing of aquaculture products. Aquaculture products play a significant role in helping countries handle the issue of food insecurity. The aquaculture sector plays an important role in global markets more so in the production of sea food. This has caused this sector to be subject to food safety mechanisms and controls. In some developed countries, there are statutory hazard analysis critical control point methodologies. This is very crucial more so with the international sale of aquaculture products among countries. Among most countries at the international level, there are trade and safety regulations agreed upon by the countries (Hossain et al., 2020).

**Figure 10.1**: There are a variety of aquaculture products sold in both local and international markets.

*Source: https://www.foodingredientsfirst.com/news/a-growing-appetite-for-fish-fao-highlights-methods-to-boost-sustainable-aquaculture.html*

There are various management practices vital in the making of aquaculture products. Some of the practices considered as integral components include traceability and risk assessment. With the increasing demand for aquaculture products some developed countries have experienced a 50% increase in seafood export through global trade. Most of the aquaculture products being exported include mollusks, salmon and shrimp among others. As the aquaculture sector continues to expand, more products are being exported to other countries. It is expected that for aquaculture's development will be considered long-term viability as it will be market-driven. This will see the sector take into consideration consumer demand. The sector is expected to be able to reach a capacity where it can adapt to the legislative and structural demands of its targeted markets.

However, the success of the aquaculture sector can be affected by some externalities. These externalities are detrimental in the manner in which the sector will achieve its objectives. This involves the sector handling issues such as sustainability, interactions and traceability with the surrounding. Schemes of best practice should be developed and they should address various issues. The aquaculture sectors should work hand in hand with other production sectors so that they can successfully meet their goals. With the marketing and sale of aquaculture products, there is a need for relevant stakeholders in the aquaculture sector to avoid discriminatory tariffs with regards to trade. Most developed countries have eliminated discriminatory tariffs to boost the importance of aquaculture being a major export-earner. This should be emphasized as the aquaculture sector plays a major role in ensuring food security.

**Figure 10.2**: Developments made to the aquaculture sector have resulted in the increase in the amount of aquaculture products produced.

*Source: https://www.ifpri.org/sites/default/files/gfpr/2015/feature_3087.html*

There are numerous developments in other sectors that affect the aquaculture sector. The aquaculture sector has been able to respond to the developments by generating a source of understanding of the complex issues experienced during the production and marketing of various food products. Aquaculture products need thrive in the very competitive product markets. It is very crucial that all the relevant bodies take responsibility for the products as well as the processes involved in making the products.

By taking responsibility for the product and the manner in which they are produced, the sector is able to change the manner in which consumers view fish as food. By changing the perception, they create an avenue where they can increase the supply for fish as well as their products from cultured sources. With the increase and expansion of the aquaculture sector, there are laws and requirements placed to guide the production of such food products so that food safety is ensured. When generating these laws, relevant stakeholders questioned the nutritional benefits of aquaculture food products (Hemaiswarya et al., 2011). This resulted in some countries requiring those producing aquaculture food products undergo certain inspection procedures each varying with the species of fish being produced. For the various fish species, different levels of hygiene are required. This is to ensure food safety for the consumers. Among the measures taken was the adoption of preventive systems known as the hazard analysis critical control point. Many developed countries have shifted their initial systems so as to adopt the HACCP system. These systems are generally linked to "prevention".

When dealing with the marketing of aquaculture products, there is a great need for international agreements. This allows easy trade among member countries. These agreements focus on food safety more so with aquaculture food products. According to recent statistics, about 40% of fish produced by countries are traded internationally. With international trade, those producing aquaculture products should be marketed according to the international laws agreed upon. These laws allow aquaculture companies to package and market their products in various countries.

To ensure that marketing is easily done, international countries met together and worked towards the harmonization of national regulations. This means that regulations and rules laid down on food protection will ensure a harmonic level of food safety. With trading and marketing of aquaculture products international guidelines, standards and recommendations are of high importance. Though there are such laws, it is quite difficult to conduct an analysis of the level of the regional and international trade among the countries known to be major producers of aquaculture products. In most of these countries, trade has not been well documented. When marketing aquaculture products. It is important that companies consider their local and international consumers. With international products, there are some products known to be popular in different countries. These products are said to be the driving force in the development of the aquaculture sector in both developed and developing countries.

## 10.2 AQUACULTURE PRODUCTS

Marketing of the different aquaculture products will vary depending on the type of products and consumers. There are certain aquaculture products known to be produced by small-scale farmers while others are produced by large-scale farmers. Most aquaculture products sold in international markets are mostly marine fish products as they are considered exotic and are highly demanded in other countries. Some of the marine products are scarcely found in other countries. Most of the marine aquaculture products are sold in international markets and few are left for local consumers. Among the aquaculture products sold in international markets include catfish, salmon, tilapia and trout. Products obtained from reptiles include alligators, turtles and crocodiles. Mollusks products include clams, oysters and mussels while crustacean products include crawfish, shrimp, lobsters and crabs.

## 10.3 CRUSTACEANS

**Figure 10.3**: Products such as shrimps are in high demand in the aquaculture market.

*Source: https://www.shutterstock.com/search/crustacean*

These kinds of aquaculture products are mostly marine though there are certain species that can be reared in freshwaters. Among the crustaceans, marine shrimps are the most commonly sold aquaculture products. They are usually traded to international markets. The numerous developments made to the aquaculture sector have resulted in shrimp being highly demanded in the international market. Statistics show an increase in the amount of shrimp traded in international markets over a period of 8 years. From the general amount of shrimp sold to international markets, about 25% is from the aquaculture sector. To stabilize the shrimp industry, farmers have mostly revolved around the farming of shrimp. However, in countries like Latin America and Asia, there has been a decline in the amount of shrimp produced. The decline is a result of the incidence of diseases. The incidence of diseases really affected the demand and supply of shrimp. This resulted in high prices of shrimp in these countries (Hasimuna et al., 2020).

The major shrimp markets include the United States and Japan. Globally, the major exporters of shrimp are India, Bangladesh, Thailand, Vietnam, Indonesia Mexico and Ecuador. Predictions in aquaculture show that in the coming years, there is an expected increase in the demand for shrimp. This will be demonstrated by the expansion of shrimp markets in countries like Malaysia, Korea, Asia, Thailand and China. This will see the expansion of local economies in those countries. The increase in the exportation of shrimp will result in the decline in the availability of shrimp for local sellers. Traditional importers will also get a challenge in accessing shrimp which may put pressure on them to increase the price of fish to compensate for the decline in fish production.

**Figure 10.4**: Shrimp is used in various restaurants in preparing certain dishes.

*Source: https://en.wikipedia.org/wiki/White_boiled_shrimp*

Shrimps when exported to international markets, they are usually sold to restaurants. There are certain species of shrimps on demand. They include white leg shrimp, giant tiger prawn, southern rough shrimp, Akamai paste shrimp, fleshy prawn, banana prawn and northern pawn. There are several uses of shrimp as food. While marketing fish, the seller should be aware of the different uses of shrimp. In most cases, when marketing shrimp, sellers usually market them based on their grade, uniformity, presentation and color. Shrimp have a lot of nutritional value more so omega-3 fatty acids. Most sellers usually market the meat of shrimp. Among the dishes prepared using shrimp is sushi. Shrimp can also be harvested and sold in aquariums.

Other than shrimp, crayfish is another crustacean marketed and sold in international markets. Supply and selling of crayfish has increased with the developments made to the aquaculture sector. There is a rise in the amount of crayfish exported to international markets. The demand for crayfish is lower than that of shrimp. Crayfish are often confused with lobsters. These crustaceans are usually reared in freshwater and have major markets in the United States. However, with crayfish, their supply can easily be affected by pollution. There is a noted trend in the increase in crayfish exportation in countries like China. This is because more farmers are adopting the farming of crayfish to meet the demand for this product in countries like Europe. The Hubei province announced that in 2015 they sold crustaceans worth 190 million dollars and about 27% of the sales were from crustaceans. This province is becoming a major crayfish exporter (Freitas et al., 2020). Currently, China is the leading producer of crayfish. There are several aquatic companies in China involved in the production of crayfish. These companies have heavily invested in the production of crayfish. Farmers working with these countries use the latest technologies in the rearing of fish.

**Figure 10.5**: Different species of crayfish are sold in international markets.

*Source: https://www.bbc.com/news/science-environment-27812367*

Other countries are following the same trend as China and are reaping benefits from exporting crayfish products. Many countries are generating large amounts of revenue from the sale of these aquatic products. Globally, the demand for crayfish is high as they are eaten worldwide. However, there are certain parts of the crayfish that are eaten. These parts are marketed. When sold to restaurants, crayfish is used in making dishes such as bisques and soups. In most cases, companies sell crawl meat. In other instances, the entire crawl is sold. This mostly applies to restaurants as there are dishes that require the entire body of the crayfish to be presented. As mentioned earlier, the United States is known to produce crayfish. However, most of the harvested crayfish are not exported but locally used. In the United States, Louisiana is known to produce a large percentage of crayfish. About 93% of crayfish farms are found in Louisiana. When marketing crayfish, producers should consider the fact that certain religions do not permit the eating of crayfish as they have both scales and fins.

Crayfish can be packaged and marketed as human food. In other cases, it is marketed as bait food for fishermen. This is because several fish species are known to prey on crayfish. When marketing it as bait, most companies market the tail though there are farmers who may opt to use the entire crayfish as bait. They usually hook the eyes in a manner that they are not damaged and allows them to be more active. In this case, crayfish are harvested, packaged and sold while alive.

Other than using crayfish as bait or food, there are those who sell them as pets kept in aquariums. Crayfish are omnivores and can therefore feed on anything resulting in them being considered as good pets. They are carefully harvested from ponds and transported to pet stores where they are stored in aquariums before they are sold to new home owners. When keeping crayfish as pets, there are certain species that can be considered. This is because there are certain species that carry diseases and can spread the disease to other crayfish. This is why countries like New Zealand, United Kingdom, Australia and the United States consider the importation of crayfish hazardous to local rivers.

**Figure 10.6**: Different methods are used in the marketing of crabs.

*Source: https://foodal.com/knowledge/how-to/crabs-ultimate-guide-to-buying-reparing-and-cooking/*

Crabs are highly on demand. Some may consider them to be on the same level as shrimps. Countries like the United States are known to export several species of crabs. Countries known to be major exporters of crabs are said to earn about 350 million dollars in a year. Among the species of crabs exported, the blue crab is highest on demand followed by the Dungeness crab. This species of crab is mostly caught in the pacific coast while others are caught in Oregon. For countries like Alaska, commercial species of fish include the snow and king crabs. Crabs are marketed and sold in different ways. There are those sold when fresh while others are frozen. In other companies, certain parts of crabs are sold such as the legs. There are those who sell fresh crab meat while others sell it while canned or pasteurized. There are different forms of crabs as products. Some crabs are sold when dead while others when alive. Others are sold as freshly cooked whole crabs while others as frozen cooked whole crabs. In the marketplace, different methods of preservation are used in sustaining crabs. There are those who keep them in freezers while others used other storage techniques (Faruk et al., 2004).

A common form of crabs as a product is when they are picked by consumers from off the shelf. Canned crab products vary depending on the species of the crab. When marketing crabs, the targeted audience is usually those who want a highly nutritious diet. Crabs usually have low fat and are rich in proteins. They also contain high amounts of minerals such as selenium. Most companies usually put nutritional labels on the crab species sold. With the noted development in the aquaculture sector, there is an increase in the production of crabs. Other countries such as China are producing large amounts of crabs and exporting them to other countries such as Hong Kong and Japan.

## 10.4 FINFISH

Finfish crown the list of the most exported aquaculture products. About 49% of finfish products are obtained from aquaculture. The main species of finfish sold are the crap species. These species are known to be mostly consumed locally in countries like India and China. Finfish aquaculture is said to be different from that of shrimp in the sense that there seems to be a high demand and a good perception for finfish products. There is also a trend among consumers where they tend to purchase convenience products of fish species that are much cheaper such as tilapia. Among the finfish species marketed and sold in international markets are as follows:

### 10.4.1 Salmon

Over the years, there has been a notable increase in the number of salmon farms. The growth is said to be rampant over the past decade. The most-traded salmon species are the Atlantic salmon followed by the coho salmon. The growth in salmon production can be mirrored by its trade. In countries like Norway, United Kingdom and Chile, there is increasing activity in the production in that their domestic markets have grown.

**Figure 10.7**: Most aquaculture farmers opt to farm salmon as it is easier.

*Source: https://solidstarts.com/foods/salmon/*

With the increase in salmon production, many farmers are embracing this aquaculture product as it is considered the ultimate farmed food. Salmon is considered the most efficient food system when compared to other meat products such as pork, beef and poultry. Salmon is considered as the best food source as it records the highest edible yield per 100g at 61 Kg. It also has a feed conversion of 1.1. Compared to other aquatic animals, the meat production is almost that of beef , three times more than that of poultry and four times that of pork. With regards to protein and energy retention, salmon is known to retain more energy compared to other animals. For instance, Salmon is known to retain about 31% of protein while beef retains about 15% of protein. More farmers opt to rear salmon compared to other aquatic animals because rearing of salmon is said to be more environment friendly (Deb et al., 2021).

The rearing of salmon is said to bear the least environmental impact compared to the rearing of animals such as poultry and beef. This is well demonstrated by the carbon footprint of salmon when compared to other animal products. The carbon footprint for salmon is 2.9 Kg of the edible product while pork and chicken have 5.9 per Kg and 2.7Kg of edible product respectively. The list is topped by that of beef as beef records a carbon equivalent of about 30 per Kg of edible product. Also, the rearing of salmon is more efficient with regards to water use. Though salmon required 2000 liters of water for every Kg of edible meat, its water intake is about half the amount consumed by chicken. This amount is about a third of the amount of water used in rearing pigs and very little compared to the amount of water used in rearing beef which is about 15400 liters.

As mentioned earlier, the major species of salmon exported to other countries is salmon. Norway is the major exporter of salmon to other countries with its main market being the EC. The Coho salmon species is known to be exported by countries such as Chile and its major markets being the United States and Japan. Chile follows Norway as the second-largest exporter of Atlantic salmon. With the growing increase in salmon production, competition among markets has also increased. This has resulted in the decline in salmon prices. The current price level of salmon renders salmon a medium-priced product among several international seafood markets.

## 10.4.2 Trout

**Figure 10.8**: Trout production and marketing is second to that of Salmon.

*Source: http://www.trout-tree.com/trout/4594505152*

When compared to salmon, trout is sold and marketed in international markets at a lower rate. Though the exports are quite good, they are still lower than that of salmon. The issue with trout production is the fact that there is split production. Some may refer to it as split production. It is split according to the portion size where there are large and small pieces of trout and split according to countries of production where Europe is known to be a major producer and exporters of trout. The portion-size fish is normally consumed in countries that produce trout. However, there are countries that have been able to modify their farming so that they produce large-sized trout for commercial purposes. These countries include Finland, Norway and Chile. They produce large-sized trout and sell them in Japanese markets. In most cases, trout is usually sold as a replacement for salmon species more so the Coho salmon. Japan being a major importer of trout has noted an increase in the amount of trout sold to other countries.

### 10.4.3 Tilapia

**Figure 10.9**: Tilapia is the most common species of finfish aquatic animals.

*Source: https://www.aquaculturealliance.org/advocate/evaluating-genetic-parameters-for-resistance-to-tilapia-lake-virus-in-nile-tilapia/*

Production of various species of tilapia had increased over the years. This has resulted in an increase in international trade though it may not be as much as traded trout may require. International trade of tilapia is still expected to increase. This scenario is evident in countries like Indonesia, China, Thailand, Japan, Colombia, Costa Rica, Ecuador and the United States. Israel, Jamaica and Zimbabwe produce tilapia and export it to EC. Trade between the two countries is described as modest. Globally, the biggest exporters of tilapia include the Taiwan province of China. It is known to supply high-quality tilapia fillets to Japan where they are sold in the sashimi market. When exported to the American market, tilapia is frozen and shipped to the countries. Tilapia production and exports are usually regulated. From the total amount of tilapia produced in the Taiwan province, 35% is sold in local markets for domestic use and the rest is exported to other countries. Of the total tilapia imports in the United States, about 80% is from the Taiwan province (Dauda, 2020).

With the numerous developments made in the aquaculture sector, more countries are entering science. Some countries are starting to export more of their farm produce. Vietnam had recently joined the aquaculture sector and will be producing tilapia for exportation. Indonesia and Thailand are known to export about 5% of their products. China exports its tilapia to the United States. Of all the aquaculture products sold all around the world, tilapia is ranked third after shrimp and salmon. Most of these products are imported by the United States. Over the years, imports made by the United States have increased. This trend is expected to continue resulting in the decline of tilapia prices as tilapia imports will increase in the United States and Europe. With Europe, the market for tilapia will have to be developed first for it to get increased imports.

### 10.4.4 Seabass and seabream

Globally, there is a high demand for marine aquaculture products. Countries like Europe are known to have large marine fish farms. In Europe, the fish farms are focus on the production of gilthead seabream and European seabass. Europe is trying to match salmon production in other countries. Over the years, seabass and seabream production has increased to the extent that it is able to export about 90% of its total products to countries such as Spain and Italy. The major seabass and seabream producer is Greece. It exports about 70% of its domestic produce. The main market for these aquaculture products is Italy. Initially, it was the exclusive market for seabass and seabream products from Greece. With the developments made to aquaculture markets, Greece has expanded its markets and is selling its aquaculture products to countries like France, United Kingdom and Germany among others. It is expected that more countries will follow the trend.

Packaging of seabass and seabream products varies according to the country. There are countries known to import live fingerlings. Croatia, Malta and Greece buy their fingerlings from France and Spain. From the year 1990-1999 production and sell of seabream has tremendously increased causing prices to drop by half. Initially, prices were $16 per Kg currently they are about $5 per Kg. There are a number of factors that have contributed to the quick saturation of markets and the significant decline in prices of such commodities (Dauda et al., 2019). Some of the factors include: the presence of small traditional markets for the products when compared to markets for Atlantic salmon, lack of proper marketing and presence of developed markets, absence of diversified products and the lack of advanced technology that could be well used to improve productivity.

Some of the technologies include efficient feeders, genetic modifiers and feeding strategies among others. The decline in product prices goes a long way in helping countries develop new markets and improve already existing markets. This can only work if there are considerable profit margins that can be sustained throughout the production process though there are numerous improvements made to improve productivity, intensify marketing efforts and diversification of products.

### 10.4.5 American catfish

In the United States, American catfish is the fifth most consumed species of fish. However, this species of fish is not exported as much. Most of the produce goes to domestic markets. Even though exports are limited, most of this species of fish is exported to Europe. Production and sale of the American catfish bear similarities to that of tilapia in terms of consumer demands and the ease in preparing fillets.

### 10.4.6 Seaweed

Over the years, there has been a notable increase in the farming and production of seaweed more so in the last decade. Currently, about 87% of produced seaweed is produced is sold. A large percentage of the produce is sold in local markets for domestic food though there is a growing trend in the amount of seaweed sold in the international market. The leading seaweed producer is China. It exports its seaweed to countries like Japan and Korea. When exported to Korea in large amounts, Korea exports some of its imports to Japan. There are two main species of seaweed exported. They are the red seaweed and brown seaweed. The Gracilaria species of seaweed is produced by Chile and is exported globally. This species of seaweed is used in manufacturing agar. Significant amounts of red seaweed are usually exported by Indonesia, the Philippines and Tanzania. They are exported to Japan, Denmark and the United States. The total amount of seaweed imported to EC is usually exported by Australia, the Philippines, Indonesia and Chile.

### 10.4.7 Mollusks

Mollusks are mostly produced for domestic purposes. This goes to show that international trade of these products is usually limited as less amounts of mollusks are exported to other countries. About 10% of the total amount of mollusks produced is sold in international markets. The Republic of Korea and China are known to be major exporters of mollusks while France, Japan and the United States are the leading importers. The low rates of mollusks exports are due to the contributions made by mollusks in trade.

With mollusks production, there are numerous products under it among them being mussels, clams and scallops. However, in international trade, clams and scallops are the most traded items. They are sold frozen or fresh. In 1998 imports of frozen and fresh scallops increased generating about $510 million after the sales. In the case of clams, the imports have increased and in the year 1997, it was valued at $257 million. However will the sale of mussels, the trend is opposite of that of clams and scallops in 1992. Imports decreased. Work was done to improve the sector so resulting in it being valued at $220 million. In 1998, oyster imports increased and were valued at $122 million (Bossier & Ekasari, 2017).

### 10.4.8 Seed supplies

There is a high demand for seeds used in the culture of aquatic organisms. This has created a great significance in the trade of such seeds in both local and international markets. Most of these seeds are gotten from aquaculture sources. However, it is quite difficult to monitor the production and sale of seed supplies. There are records of the levels of fingerlings of Mediterranean seabream and seabass at local levels. However, at the international level, the wild glass eels are the most sold aquaculture products. Currently, there is a notable increase in the amount of eel elvers imported by China. Also, there is a significant amount of eyed fertilized eggs of trout and salmon and the post larvae of different cultured shrimps. China and India are also exporting craps to other countries. With brood stock, there are limited amounts of such products depending on the type of products. It is expected that documentation of the sale of fingerlings and seeds will increase to tackle issues and challenges with regards to the spread of pathogens and the manner in which genetic material moves from one area to another. This will make traceability much easier.

## 10.5 AQUACULTURE MARKETING

With the recent developments made in the aquaculture sector, many aquaculture companies are working towards the production of market-oriented products. This enables the companies to better sell their products to many consumers. This means that the companies need to make appropriate market surveys before they can fully integrate their production program.

By conducting the surveys, aquaculture companies will be able to take into consideration the production cost, consumer acceptance, the manner in which products can be marketed and processed to a cost that is relevant. In most countries, there are workshops done to check the prices of aquaculture products in various countries. From the workshops done, it was noted that there are differences in the consumer demand in different regions. In some areas, there are those who prefer fish from freshwater while there are those who prefer indigenous species cultured in a certain manner. Marketing such products will not be as difficult as they are already in demand.

There is a challenge when the company decides to produce and market non-indigenous species that are not familiar to consumers. Sales may be low. This may require the company to promote and in some instances conduct product development. Aquaculture companies also need to go with the flow of current trends in the business world among them hiring women in different areas. Among the current trends, there is the high preference for convenient foods more so in Mexico and Cuba. In some countries, citizens are not familiar with aquaculture products. For those who like ready foods, companies may have to package and sell processed aquaculture products that are ready to cook or in a form that is ready to eat.

If the products are to be sold in an international market, company owners will have to conduct research on the international market. In this sense, foreign market information is very useful. The company also needs to look at product specifications and the quality level standards placed by the country for a given product. These data should be updated at a certain frequency. For certain regions, it is very beneficial if aquaculture products are added to the contemplated Regional Fish Market Service. This will be very beneficial for the given region. It may be quite difficult for companies to introduce new aquaculture products to the markets. The introduction of new products will require several preparations to be made before they are fully accepted by consumers. When dealing with the marketing of products, it is crucial for relevant bodies to put in place special marketing arrangements to be put in place. This is crucial for both large and small-scale farmers. This is because their products will be sold locally or internationally.

To better understand by many aquaculture companies invest lots of money in marketing, we look at the benefits of marketing aquaculture products.

## 10.6 BENEFITS OF AQUACULTURE MARKETING PRODUCTS

The first benefit of marketing aquaculture products is that it helps companies determine the price for various aquaculture products. Marketing ensures that aquaculture farmers are able to get income from the sale of aquaculture products. Through marketing, farmers and marketers are able to get back the money they spent due to floating and fixed capital throughout the purchasing and production process. By marketing their aquaculture products farmers and marketers avoid glut on location and transportation of the products from one area that has a shortage of the products. By doing so, farmers are encouraged to expand and sustain their production to the highest achievable level. Through marketing, both farmers and marketers are able to know the preference of their consumers. This way they know the kind of aquaculture products to produce and sell. During the research process, farmers and marketers conduct research on the species and types of fish preferred by consumers. Marketing also creates an avenue for aquaculture products to be exported to foreign countries allowing them to get foreign exchange. This also gives the nation and farmers conservation. It also fosters international relationships among countries of the world.

Marketing is essential in that it is useful in determining the manner in which aquaculture products should be processed, sorted, stored, graded and sold to customers. Research on marketing methods enables farmers to know the type of packaging that will aquaculture products seem more appealing, appetizing and attractive to consumers so that the products sell. Marketing also creates avenues for employment. Some individuals may be employed as marketers or researchers.

## 10.7 FEATURES OF MARKETING OF AQUACULTURE PRODUCTS

When marketing aquaculture products, there are characteristics common in all forms of marketing. Some of the common marketing features are as follows:

Production begins at the lowest level which is of individual farmers. At this level, farmers are referred to as producers. At this level, farmers rear and harvest aquaculture products such as fish from water bodies. The harvested products undergo processing to a level that will render them fit for markets. The final level of the production level is the consumer. They are the ones who acquire finished aquaculture products. Consumers decide whether to eat or change the form of the product to another that is entirely different from the original product. For example, there are areas where fish is consumed after oil was extracted from it. Usually, there are several producers throughout the world. Previously, producers were found in areas near water bodies and were found throughout the society. Over the years, ponds, cages and tanks increased all over the world increasing the production of aquaculture products.

When dealing with aquaculture products, it is crucial for one to understand that most of these products are usually perishable and supply can be limited to the season experienced. Even though most of these aquaculture products will undergo processing and preservation by using various methods, these aquaculture products can remain viable for certain duration of time. If extended, these products will lose their nutritional value (Benetti et al., 2021). They also get wasted and spoiled. According to the season, the supply of most of the aquaculture products is limited in certain seasons. It is not the same throughout the year. This means that there are times when these products will be in large amounts and times when the supply will be limited.

Among the challenge with the production of aquaculture products is that there is limited support from institutions. Studies show that institutional support in the supply of aquaculture products to various markets is very small to make a significant improvement for such markets. The amount of aquaculture products shipped to local and international markets by institutions and organizations is very limited. Usually, large amounts of these products are supplied by farmers. There are those who work with some organizations in the production and marketing of these products.

## 10.8 MARKETING CHANNELS

When marketing products, it is important for farmers to know the different available marketing channels. With domestic markets and distribution of aquaculture products, the area is dominated by several intermediaries. Aquaculture products traded in the international market usually go through private channels before they are exported. With the distribution of aquaculture products, there are four main levels of product distribution to markets.

The first level is made of primary markets. Most of these markets are located in villages, crossroads or district headquarters. Most of these markets are found in areas where aquaculture products are caught. Usually, farmers bring the different products harvested to the primary markets. This means that most of these markets are dominated by aquaculture products captured in open water sources or ponds. In rural areas, these markets can be held twice or thrice a week. In other areas, these markets are held daily. Some of these markets run during morning-hours and are known to be attended by fewer numbers of buyers and sellers.

After the primary markets, there are secondary markets. This is an advancement from primary markets in that products are moved from farmers or primary markets and taken to other areas close to the road, rails or wholesalers who will sell them in further areas. From these markets, aquaculture products are distributed through different channels. In most cases, the products are taken to urban markets by commission agents. These agents avail the products to wholesalers and other buyers (Bondad-Reantaso et al., 2005). After the secondary markets are the higher secondary markets. Aquaculture products from secondary markets are taken to higher secondary markets that may be referred to as terminal markets. In such markets, there are one or more wholesale centers or markets. These markets are known to have good connections to rail, rivers and roads. In some cases, the higher secondary markets are made up of a combination of several secondary markets. Most of these markets handle large amounts of both fresh and salty water aquatic animals.

# References


1. Akter, M. N., Parvez, I., &Patwary, Z. P. (2016).Beneficial effects of probiotics in aquaculture. *International Journal of Fisheries and Aquatic Studies*, *4*(5), 494-499.
2. Anderson, J. L., Asche, F., &Garlock, T. (2019).Economics of aquaculture policy and regulation. *Annual Review of Resource Economics*, *11*, 101-123.
3. Amenyogbe, E., Chen, G., Wang, Z., Lin, M., Lu, X., Atujona, D., &Abarike, E. D. (2018).A review of Ghana's aquaculture industry. *J Aquac Res Development*, *9*(545), 2.
4. Assefa, A., &Abunna, F. (2018). Maintenance of fish health in aquaculture: review of epidemiological approaches for prevention and control of infectious disease of fish. *Veterinary medicine international*, *2018*.
5. Balcázar, J. L., & Ma, S. (2007). Probiotics as control agents in aquaculture. *Journal of Ocean University of China*, *6*(1), 76-79.
6. Barman, D., Nen, P., Mandal, S. C., & Kumar, V. (2013). Immunostimulants for aquaculture health management. *J Marine Sci Res Dev*, *3*(3), 1-11.
7. Benetti, D. D., Suarez, J., Camperio, J., Hoenig, R. H., Tudela, C. E., Daugherty, Z., ... & Stieglitz, J. D. (2021). A review on cobia, Rachycentroncanadum, aquaculture. *Journal of the World Aquaculture Society*.
8. Bondad-Reantaso, M. G., Subasinghe, R. P., Arthur, J. R., Ogawa, K., Chinabut, S., Adlard, R., ...&Shariff, M. (2005). Disease and health management in Asian aquaculture. *Veterinary parasitology*, *132*(3-4), 249-272.

9. Corsin, F., Mohan, C. V., Padiyar, A., Yamamoto, K., Chanratchakool, P., & Phillips, M. J. (2008). Codes of practice and better management: a solution for shrimp health management. *Diseases in Asian Aquaculture VI*, 419-432.
10. Bostock, J., McAndrew, B., Richards, R., Jauncey, K., Telfer, T., Lorenzen, K., ...& Corner, R. (2010). Aquaculture: global status and trends. *Philosophical Transactions of the Royal Society B: Biological Sciences*, *365*(1554), 2897-2912.
11. Bossier, P., &Ekasari, J. (2017).Biofloc technology application in aquaculture to support sustainable development goals. *Microbial biotechnology*, *10*(5), 1012-1016.
12. Dauda, A. B., Ajadi, A., Tola-Fabunmi, A. S., &Akinwole, A. O. (2019). Waste production in aquaculture: Sources, components and managements in different culture systems. *Aquaculture and Fisheries*, *4*(3), 81-88.
13. Dauda, A. B. (2020). Biofloc technology: a review on the microbial interactions, operational parameters and implications to disease and health management of cultured aquatic animals. *Reviews in Aquaculture*, *12*(2), 1193-1210.
14. Deb, A., Chandan, C. S. S., Roy, P., Hossain, M. I., & Bari, S. M. (2021). Inland Aquaculture and Fish Health Management: A Case Study of Sylhet District in Bangladesh. *Aquaculture Studies*, *21*(3), 129-137.
15. Faruk, M. A. R., Alam, M. J., Sarker, M. M. R., &Kabir, M. B. (2004). Status of fish disease and health management practices in rural freshwater aquaculture of Bangladesh. *Pakistan Journal of Biological Science*, *7*(12), 2092-2098.
16. Freitas, J., Vaz-Pires, P., &Câmara, J. S. (2020). From aquaculture production to consumption: Freshness, safety, traceability and authentication, the four pillars of quality. *Aquaculture*, *518*, 734857.
17. Hasimuna, O. J., Maulu, S., &Mphande, J. (2020). Aquaculture Health Management Practices in Zambia: Status, Challenges and Proposed Biosecurity Measures. *Journal of Aquaculture Research and Development*, *11*(3), 1-6.
18. Hemaiswarya, S., Raja, R., Kumar, R. R., Ganesan, V., &Anbazhagan, C. (2011). Microalgae: a sustainable feed source for aquaculture. *World Journal of Microbiology and Biotechnology*, *27*(8), 1737-1746.

19. Hossain, M. S., Koshio, S., &Kestemont, P. (2020). Recent advances of nucleotide nutrition research in aquaculture: a review. *Reviews in Aquaculture*, *12*(2), 1028-1053.
20. Jana, P., Karmakar, S., Roy, U., Paul, M., &Bera, A. K. S. K. K. (2018). Phytobiotics in aquaculture health management: A review. *Journal of Entomology and Zoology Studies*, *6*(4), 1422-1429.
21. Kalbassi, M. R., Abdollahzadeh, E., &Salari-Joo, H. (2013).A review on aquaculture development in Iran. *Ecopersia*, *1*(2), 159-178.
22. Labh, S. N., &Shakya, S. R. (2014).Application of immunostimulants as an alternative to vaccines for health management in aquaculture. *International Journal of Fisheries and Aquatic Studies*, *2*(1), 153-156.
23. Li, X., Li, J., Wang, Y., Fu, L., Fu, Y., Li, B., & Jiao, B. (2011). Aquaculture industry in China: current state, challenges, and outlook. *Reviews in Fisheries Science*, *19*(3), 187-200.
24. Liao, I. C., & Chao, N. H. (2009). Aquaculture and food crisis: opportunities and constraints. *Asia Pacific journal of clinical nutrition*, *18*(4), 564-569.
25. Luis, A. I. S., Campos, E. V. R., de Oliveira, J. L., &Fraceto, L. F. (2019). Trends in aquaculture sciences: from now to use of nanotechnology for disease control. *Reviews in Aquaculture*, *11*(1), 119-132.
26. Lulijwa, R., Rupia, E. J., & Alfaro, A. C. (2020). Antibiotic use in aquaculture, policies and regulation, health and environmental risks: a review of the top 15 major producers. *Reviews in Aquaculture*, *12*(2), 640-663.
27. Marques, F. B., Watterson, A., da Rocha, A. F., &Cavalli, L. S. (2020).Overview of Brazilian aquaculture production. *Aquaculture Research*, *51*(12), 4838-4845.
28. Michael, E. T., Amos, S. O., &Hussaini, L. T. (2014).A review on probiotics application in aquaculture. *Fisheries and Aquaculture Journal*, *5*(4), 1.
29. Mishra, S. S., Rakesh, D., Dhiman, M., Choudhary, P., Debbarma, J., Sahoo, S. N., & Mishra, C. K. (2017). Present status of fish disease management in freshwater aquaculture in India: state-of-the-art-review. *Journal of Aquaculture & Fisheries*, *1*(003).
30. Mohapatra, S., Chakraborty, T., Kumar, V., DeBoeck, G., &Mohanta, K. N. (2013). Aquaculture and stress management: a review of probiotic

intervention. *Journal of animal physiology and animal nutrition*, *97*(3), 405-430.

31. Muhala, V., Rumieque, A., &Hasimuna, O. J. (2021). Aquaculture production in Mozambique: Approaches and practices by farmers in Gaza province. *The Egyptian Journal of Aquatic Research*, *47*(1), 87-92.
32. Narbonne, J. A., Radke, B. R., Price, D., Hanington, P. C., Babujee, A., & Otto, S. J. (2021). Antimicrobial Use Surveillance Indicators for Finfish Aquaculture Production: A Review. *Frontiers in veterinary science*, *8*, 175.
33. Ngajilo, D., &Jeebhay, M. F. (2019).Occupational injuries and diseases in aquaculture–a review of literature. *Aquaculture*, *507*, 40-55.
34. Otta, S. K., & HOQUE, F. (2020). Health Management in Coastal Aquaculture System. *Journal of the Indian Society of Coastal Agricultural Research*, *38*(2).
35. Opiyo, M. A., Marijani, E., Muendo, P., Odede, R., Leschen, W., &Charo-Karisa, H. (2018). A review of aquaculture production and health management practices of farmed fish in Kenya. *International journal of veterinary science and medicine*, *6*(2), 141-148.
36. Ottinger, M., Clauss, K., &Kuenzer, C. (2016). Aquaculture: Relevance, distribution, impacts and spatial assessments–A review. *Ocean & Coastal Management*, *119*, 244-266.
37. Panigrahi, A., & Azad, I. S. (2007). Microbial intervention for better fish health in aquaculture: the Indian scenario. *Fish physiology and biochemistry*, *33*(4), 429-440.
38. Pandey, P. K., Bharti, V., & Kumar, K. (2014).Biofilm in aquaculture production. *African Journal of Microbiology Research*, *8*(13), 1434-1443.
39. Perumal, S., Thirunavukkarasu, A. R., &Pachiappan, P. (Eds.). (2015). *Advances in marine and brackishwater aquaculture*. Springer India.
40. Quesada, S. P., Paschoal, J. A. R., & Reyes, F. G. R. (2013). Considerations on the aquaculture development and on the use of veterinary drugs: special issue for fluoroquinolones—a review. *Journal of food science*, *78*(9), R1321-R1333.

41. Raja, R. A., &Jithendran, K. P. (2015).Aquaculture disease diagnosis and health management.In *Advances in marine and brackishwater aquaculture* (pp. 247-255).Springer, New Delhi.

42. Rico, A., Jacobs, R., Van den Brink, P. J., &Tello, A. (2017).A probabilistic approach to assess antibiotic resistance development risks in environmental compartments and its application to an intensive aquaculture production scenario. *Environmental Pollution*, *231*, 918-928.

43. Rosa, R., Marques, A., &Nunes, M. L. (2012).Impact of climate change in Mediterranean aquaculture. *Reviews in Aquaculture*, *4*(3), 163-177.

44. Rodrigues, P. M., Silva, T. S., Dias, J., &Jessen, F. (2012). Proteomics in aquaculture: applications and trends. *Journal of proteomics*, *75*(14), 4325-4345.

45. Roubach, R., Menezes, A., Oh, K., &Dabbadie, L. (2019).Towards guidelines on sustainable aquaculture. *FAO Aquaculture Newsletter*, (60), 55-56.

46. Rahman, M. Z. (2014). Aquaculture drugs used for fish and shellfish health management in the Southwestern Bangladesh. *Asian J. Biol. Sci*, *7*, 225-232.

47. Rimmer, M. A., Sugama, K., Rakhmawati, D., Rofiq, R., &Habgood, R. H. (2013).A review and SWOT analysis of aquaculture development in Indonesia. *Reviews in Aquaculture*, *5*(4), 255-279.

48. Stentiford, G. D., Bateman, I. J., Hinchliffe, S. J., Bass, D., Hartnell, R., Santos, E. M., ... & Tyler, C. R. (2020). Sustainable aquaculture through the One Health lens. *Nature Food*, *1*(8), 468-474.

49. Subasinghe, R. (2017). World aquaculture 2015: a brief overview. *FAO Fisheries and Aquaculture Report*, (1140).

50. Sivasankar, P., John, K. R., George, M. R., Anushalini, S. V., Kaviarasu, D., &Petchimuthu, M. (2017). Prophylactics in shrimp aquaculture health management: A review. *Journal of Entomology and Zoology Studies*, *5*(4), 1049-1055.

51. Tacon, A. G., &Halwart, M. (2007). Cage aquaculture: a global overview. *FAO Fisheries Technical Paper*, *498*, 3.

52. Tucker, C. S., & Hargreaves, J. A. (Eds.).(2009). *Environmental best management practices for aquaculture*. John Wiley & Sons.

53. Waagbø, R., &Remø, S. C. (2020). Functional diets in fish health management. In *Aquaculture Health Management* (pp. 187-234). Academic Press.

# INDEX


**A**

Aerobic bacteria 209
Agricultural effluents 195
Airborne organisms 28
Alien species 48
Anaerobic bacteria 209
Animal welfare 9
Antibiotics agents 81
Aqua-bounty technologies 102
Aquaculture 2
Aquaculture activities 162, 164, 169, 173, 177, 185, 186
Aquaculture business 59
Aquaculture certification council (ACC) 158
Aquaculture facilities 33
Aquaculture growth rates 192
Aquaculture legislation 165, 166, 175, 182, 185
Aquaculture marine zone (MAZ) 184
Aquaculture market 193
Aquaculture producers 164, 184
Aquaculture production 64
Aquaculture production activities 125
Aquaculture productivity 69
Aquaculture products 222
Aquaculture project 185
Aquaculture rules 189
Aquaculture sector 32, 214
Aquaculture-specific legislation 165
Aquaculture ventures 186
Aquaponics 15
Aquaponics system 3
Aquatic animals 36
Aquatic environments 67
Aquatic health industry 162
Aquatic organisms 81, 236
Aquatic plants 2
Aquatic products 64
Aquatic resources 67
Atlantic salmon 8
Atmospheric oxygen 116

B

Bacteria and cultured organisms 215
Bentonite clay 6
Biomass 209
Bioscience 199
Bluefin tuna farming 74
Breeding 2, 7
Breeding businesses 203
Breeding mechanism 34
Breeding season 35
Brood fish 130
Broodstock 33, 34, 35, 36
Brush park fishing 53

C

Canned crab products 229
Capture-based aquaculture 69
Carnivorous 11, 12
Cassic animal farming methods 199
Catch-based aquaculture 68
Catch-based aquaculture (CBA) 64
Catfish 5, 6
CBA development 83
Cereal-based supplementary meals 197
Clay soils 94
Coastal lagoons 53
Coldwater fish 211
Commercial aquaculture species 76
Commercial farming operation 146
Commercial fishing 2
Commercially produced feed 114
Competitive strategy 205
Composite fish culture 14
Computer-controlled automatic feeders 152
Consumer demand 223, 237
Contaminated fish 132
Crayfish products 228
Crustaceans 16
Cultivation facility 83
Cultivation systems 61
Cultural deterioration 160

D

Denitrification 15
Disease-causing bacteria 211
Disease monitoring 165
Dissolved oxygen 211
Dissolved substances 117
DNA vaccine 204
Domestication opportunities 64
Domestic markets 235

E

Ecological repercussions 173
Economic development 64
Economic expansion 192
Economic feed conversion ratios (eFCR) 144
Economic mechanisms 189
Ecosystem 45, 114
Eel aquaculture 76
Egg incubation 51
Endangered species 44
Environmental degradation 106
Environmental impact assessment 183
Export quality control 193

F

Facultative bacteria 209
Farming facilities 79
Farming processes 28
Farming system 91
Farm-made feeds 139

Feed conversion ratio (FCR) 115
Feed expenditures 155
Feeding habits 34
Feed ingredients 140, 141, 142, 143, 147, 158
Feed management 149
Feed regulations 158
Feed selection 137
Fertilization process 91
Finfish aquaculture 230
Finfish crown 230
Fin-fish species 64
Fingerling catfish 7
Fin rot 104
Fish culture systems 90
Fishery law 164
Fish farming 2
Fishing industry 8, 43
Fish oils 9
Fish productivity 160
Fish stocking 31, 40, 44, 46
Flocculation 28
Floodplain management 56
Food insecurity 95
Food methods 153
Food safety mechanisms 222
Food stability 108
Foreign market information 237
Freshwater cultivation 68
Fry farming 14

G

Gamete production 34
Generative fertilization 37
Genetic material 236
Genetic modification techniques 33
Global aquaculture production 32
Gonadal development 34
Governmental guidelines 220
Government policies 220

H

HACCP system 224
Hatchery-reared groupers 70
Herbivore animals 202
Humane methods 15
Human safety 81
Human supervision 93
Hydroponic bed 99
Hydroxide component 115
Hypernutrification 116

I

Income-generating activities 137
Inflammation problems 10
Inhumane methods 15
Inland fishing 172
Integrated recycling systems 98
Intensive aquaculture 90, 93, 94, 117
Intensive aquaculture farming 93
Intensive fish farms 101, 104, 106, 107, 111
Intensive production systems 90
International cooperation 66
International market chains 194
International markets 100
International seafood markets 231
International trade 233

K

Kidney diseases 10

L

Lagoon cultivation 52
Land-based aquaculture 109
Land-based cattle 31
Large-scale farmers 111

Lecithotrophy 38
Leisure activities 172
Licensing system 176
Lobster farming 17
Long-term genetic integrity 130

**M**

Managerial skills 114
Marginal profitability 154
Mariculture 30
Marine aquaculture 180
Marine aquaculture aone (MAZ) 183
Marine predatory creatures 48
Marine ranching 50
Maritime tenure systems 186
Market demands 75
Market-oriented products 236
Market requirements 68
Mediterranean seabream 236
Microorganisms 13
Molecular oxygen 209
Mollusks products 225
Mortality data 78
Murray cod 122, 123, 124, 125, 126, 127, 128, 129, 130
Mussels 20

**N**

Natural aquatic populations 67
Natural environment 10, 20
Natural fisheries management 178
Natural habitats 44
Natural productivity 136, 150
Neurological necrosis virus 204
Non-indigenous species 237
Non-natural environment 121
Nutrient pollution 2
Nutrient quality 152
Nutrient-rich foods 9
Nutritional benefits 96
Nutritional consistency 38
Nutritious cuisines 10

**O**

Ocean ranching 30
Off-farm activities 201
Offshore cage technology 204
Offshore cultivation 12
On-farm feed management procedures 136
Organic fertilizer 117
Organic matter 209, 210
Ornamental fish 70
Oxygen concentrations 106, 107
Oxygen saturation 35
Ozonation 14

**P**

Pacific salmon 8, 13
Packaging 234
Paddlewheel aerators 7
Pathological effects 128
Pesticides 109
Phytoplankton 210
Pisciculture 2
Piscine tuberculosis 104
Polyculture ponds 117
Pond fish production 131
Pond systems 195
Pre-aquaculture mortality 80
Proactive management process 218
Production cycle 137
Production parameters 136
Production system 137
Progressive offshore 215
Public aquariums 30

## Q

Quadruple productivity 196
Quality assurance 180, 182

## R

Recirculated aquaculture systems (RAS) 204
Recycling systems 98, 99, 111
Regulatory framework 72
Reproductive biodiversity 78
Reproductive cells 36
Routine maintenance practices 92

## S

Safety guarantees systems 182
Salmon culture 50
Salmon farming production 8
Salmon species 13
Salt water environments 9
Saltwater fish 10
Saprolegnia fungus 105
Sea food 222
Seafood aquaculture 20
Secondary markets 240
Self-sustaining business 5
Semi-intensive fish farming 118
Semi-intensive systems 118
Severe disease 196
Shellfish hatchery production 38
Shrimp feeding activity 149
Skilled individuals 92, 111
Snvironmental repercussions 196
Social disagreements 188
Socio-economic factors 208
Soft-shelled crabs 19
Solar energy 115
Sustainable aquaculture business 163
Sustainable aquaculture production methods 66
Sustainable food systems 9

## T

Targeted farmers 65
Technological advancements 126
Tetrapods 30
Tilapia farming 3, 4
Tilapia waste 3
Total factor productivity (TFPs) 201
Toxic wastes 94
Trout production 232
Trouts 39
Tuna farming 10, 74

## U

Urban farming 99
Urban setting 101

## V

Valliculture 53
Vegetable farming 3
Victorian aquaculture 122

## W

Waste discharges 206
Water aquatic animals 240
Water clarity 121
Water law 164, 165
Water purification systems 98, 106
Water replacement 114
Water retention 95

## Y

Yeast-producing virus 204
Yellowtails market 76